ROMANTIC WRITING AND

Also by Robin Jarvis

WORDSWORTH, MILTON AND THE THEORY OF POETIC RELATIONS

REVIEWING ROMANTICISM (*with Philip W. Martin*)

Romantic Writing and Pedestrian Travel

Robin Jarvis
Principal Lecturer in English
University of the West of England, Bristol

Published in Great Britain by
MACMILLAN PRESS LTD
Houndmills, Basingstoke, Hampshire RG21 6XS and London
Companies and representatives throughout the world

A catalogue record for this book is available from the British Library.

ISBN 0–333–65814–0 hardcover
ISBN 0–333–79460–5 paperback

 Published in the United States of America by
ST. MARTIN'S PRESS, INC.,
Scholarly and Reference Division,
175 Fifth Avenue, New York, N.Y. 10010

ISBN 0–312–17531–0 clothbound
ISBN 0–312–22849–X paperback

The Library of Congress has cataloged the hardcover edition as follows:
Jarvis, Robin, 1956–
Romantic writing and pedestrian travel / Robin Jarvis.
p. cm.
Includes bibliographical references and index.
ISBN 0–312–17531–0 (cloth)
1. English literature—19th century—History and criticism.
2. Walking—Great Britain—History—19th century. 3. Travel—Great
Britain—History—19th century. 4. Romanticism—Great Britain.
5. Walking in literature. 6. Travel in literature. I. Title.
PR408.W35J37 1997
820.9'355—dc21   97–5874
                                                           CIP

© Robin Jarvis 1997, 2000

First edition 1997
Reprinted 2000

All rights reserved. No reproduction, copy or transmission of this publication may be made without written permission.

No paragraph of this publication may be reproduced, copied or transmitted save with written permission or in accordance with the provisions of the Copyright, Designs and Patents Act 1988, or under the terms of any licence permitting limited copying issued by the Copyright Licensing Agency, 90 Tottenham Court Road, London W1P 9HE.

Any person who does any unauthorised act in relation to this publication may be liable to criminal prosecution and civil claims for damages.

The author has asserted his right to be identified as the author of this work in accordance with the Copyright, Designs and Patents Act 1988.

This book is printed on paper suitable for recycling and made from fully managed and sustained forest sources.

10 9   8   7   6   5   4   3   2   1
09 08 07 06 05 04 03 02 01 00

Printed and bound in Great Britain by Antony Rowe Ltd, Chippenham, Wiltshire

For William

You cannot travel on the Path before you have become the Path itself.
Gautama Buddha

Contents

Preface to the 1999 Reprint		viii
Preface and Acknowledgements		ix
1	The Rise of Pedestrianism	1
2	An Anatomy of the Pedestrian Traveller	29
3	Pedestrianism and Peripatetic Form	62
4	William Wordsworth: Pedestrian Poet	89
5	'Indolence Capable of Energies': Coleridge the Walker	126
6	Gender, Class and Walking: Dorothy Wordsworth and John Clare	155
7	Walking and Talking: Late-Romantic Voices	192
Notes		216
Bibliography		234
Index		243

Preface to the 1999 Reprint

A small number of proof errors and points of fact have been corrected in this paperback edition, which I am pleased will increase the accessibility of *Romantic Writing and Pedestrian Travel* for scholars, students and literary walkers. Early responses to the book have reassured me that the topic is one which holds wide interest, and also underlined that every reader will want to annotate it with references to writers or texts I have neglected. I look forward to reading reviews from North America, where, as I became forcibly aware when talking on the theme at a conference in Minneapolis in November 1997, the cultural meanings of walking are very different from what they are in the UK.

R.J.
January 1999

Preface and Acknowledgements

This book had its origin many years ago in a somewhat romantic desire to engineer a fusion of my academic and personal interests. A growing fascination with how many of William Wordsworth's early poems were grounded in the experience of walking, or featured pedestrian speakers or characters, gave the first clear sign of how this might be achieved. From there, I began to research thoroughly the phenomenon of the cult of pedestrian touring which appeared in the last quarter of the eighteenth century, before returning, with an altered perspective, to Wordsworth and to other major figures in Romantic literature. The issues that now preoccupied me with these writers were diverse, but my pursuit of them had behind it the pressure of a more general question: if, as I was to find Leslie Stephen saying in 1902 ('In Praise of Walking'), the 'literary movement at the end of the eighteenth century was... due in great part, if not mainly, to the renewed practice of walking', why was this striking fact of literary history so little recognised, and why did it not shape the teaching of Romanticism in universities here and abroad? My reasoned reply to this question, and, more importantly, my own effort of redress, lie in the chapters that follow; here I shall merely gesture at a history of academic collusion with the social and intellectual prejudice with which walking, and the mentalities and aesthetic practices it nourishes, have until quite recently been treated. That 'pedestrian' passes unnoticed as a term of depreciation is no accident. The revaluation of Romantic writing I attempt in this book involves challenging a critical tradition which construes its highest achievements as those in which the body is laid asleep, and insight accrues to the motionless 'living soul'; instead, the creativity of Romantic verbal art is repeatedly referred to the conditions, qualities and rhythms of a body in motion, a travelling self making excited passage over the land, or through the streets, discovering locomotive and representational freedoms that were unavailable to previous generations.

Since I started thinking on this topic, others have belatedly initiated the scholarly debate. In particular, Jeffrey Robinson's *The Walk* (1989), Roger Gilbert's *Walks in the World* (1991), and

Anne D. Wallace's *Walking, Literature, and English Culture* (1993) have given me much to think about, as well as remitting the isolation of research in a seemingly unvisited area. I hope that these authors, should they read the present book, will consider that I have contributed something to the project they have placed on such a sure footing. I regret that Celeste Langan's *Romantic Vagrancy* (1995) arrived too late for me to be able to address its provocative thesis. For all their substantial contribution to critical enquiry, one questionable feature of these books, taken together, in their bearing on the Romantic period, is the exclusive effect of their preoccupation with the work of William Wordsworth. In a long view, Wordsworth was undeniably the most influential writer of his age with respect to the subsequent literature and philosophy of walking, but our understanding of the cultural history of Romanticism is not helped by an over-concentration on his monumental pedestrian *oeuvre*, and the present book is structured with this in mind.

I am grateful to the Research Committee of the Faculty of Humanities at the University of the West of England (UWE) for the half-year's leave in 1994 which finally allowed me to make substantial progress on this book, and to the staff of the St Matthias Library for their efficiency and courtesy. Geoff Channon has been unfailingly supportive during my time at UWE. Kate Fullbrook has taken an enthusiastic interest in this undertaking from the start, and provided an inspiring example of how intellectual life can be sustained under the weight of more mundane professional demands. I have benefited from many conversations with Bill Greenslade whilst perambulating the pubs of north Bristol. Nicholas Roe and John Goodridge have been generous with their knowledge, and I have received valuable suggestions along the way from Nora Crook, Peter Kitson, Jim Lewis, Philip Martin, Jeanne Moskal and Katherine Turner, among others. I drew much-needed encouragement in my research from the trust placed in the project by Charmian Hearne at Macmillan. Sincere thanks to Tony Pinkney for more than twenty years of friendship and intellectual companionship. To my wife, Carol, who has walked with me among the world's highest mountains, I am grateful for being with me always on the lowlier path of daily life. I am conscious of many blessings as I enter my fortieth year, and my dedication expresses the greatest of these.

R.J.
Bristol

1
The Rise of Pedestrianism

One day, during the progress of Mr. Stokes's feat, a rustic was heard to exclaim, but with no greater asperity of manner than might have been produced by the fatigue of his expedition, – 'I be a comm'd a matter o' aightean miles to zee thicky theng caal'd a PEE-DES-TREE-UN; an aater aal, I onnly zeed *a Mon a waalkin*!'

The amusing quotation above is taken from an account of an athletic feat undertaken in 1815. In the twenty days from 20 November to 9 December, John Stokes, who had begun pedestrian excursions to counter a tendency to 'excessive corpulency', walked a total of 1000 miles, covering the ground at a rate of 50 miles in 12 hours each day. The Editor of the *Bristol Journal* referred to his accomplishment as 'the climax of what this age of Pedestrianism has afforded'.[1]

Stokes's walk was one of many such performances, stressing either speed or endurance, that were popular around this time , as the allusion to an 'age of Pedestrianism' suggests. However, though there may have been a rage for walking, the anonymous rustic's exclamation carries the implication that the concept of pedestrianism, and the word itself, were still fairly recent, or at any rate had not percolated all levels of society. The *Oxford English Dictionary* lends some support to this hypothesis. The first recorded instance of the adjectival form, 'pedestrian', in the literal sense I am concerned with here ('On foot, going or walking on foot; performed on foot; of or pertaining to walking'), comes from a letter written by Wordsworth to his Cambridge friend William Mathews in August 1791. The first use of the cognate noun form occurs in *The Observant Pedestrian*, a kind of sentimental journey on foot published in 1793, while the *OED*'s editors can find no occurrence of the noun 'pedestrianism' prior to *The Sporting Magazine* in 1809. Interestingly, the pejorative metaphorical sense of 'pedestrian' ('Applied to plain prose as opposed to verse, or to verse of prosaic character; hence, prosaic, commonplace, dull, uninspired') has a longer recorded history, and the age and persistence of this usage have implications that I shall be exploring later.

A study of successive editions of Samuel Johnson's *Dictionary of the English Language* provides further convincing evidence of how the new word and idea entered general use in the early Romantic period. The first edition of 1755, subsequent editions revised by Johnson himself, and posthumous revisions including the eighth edition of 1799, include only the related but limited adjective 'pedestrious', which has the specialised use of designating flightless creatures. It is only in the more comprehensively revised edition of the *Dictionary* produced by H.J. Todd in 1818 that 'pedestrian' is belatedly recognised, both in its adjectival ('On foot') and nominal ('One who makes a journey on foot; one distinguished for his powers of walking') senses; the nominal form is tellingly marked as 'modern'.

Of course, as the mention of 'pedestrious' indicates, the language was not without resources to express those meanings related to walking which it had need of prior to the Romantic period. The *OED* actually records another obsolete form, 'pedestrial', that was not included in Johnson, and which carries the primary meaning of 'going on foot', as it does in the Elizabethan travel narrative, *Coryats Crudities* (1611). But the appearance of the new forms towards the end of the eighteenth century, especially the nominal forms which have the effect of granting social identity and recognition to one who practises this particular mode of travel, together with the contemporaneous coining of a new verb, 'pedestrianize', all suggest that the semantic values of the superseded words were quite different, and that the language is adapting at this time to reflect a culturally significant development. That this is so, and that the 1780s and 1790s are the crucial transitional period in which one can observe the social and ideological meanings of walking being contested and redefined, is what I shall argue in this chapter.

What is the explanation for the linguistic ripple I have described? What kind of walking was taking place at the end of the eighteenth century? On what kind of scale? Where, and with what objectives in mind? How did all this mark a departure from antecedent norms?

To begin with, there were the athletic pedestrians represented by the corpulent John Stokes. Walter Thom's *Pedestrianism*, published in 1813, is a study of these remarkable and often eccentric individuals. Lexicography credits Thom with introducing 'pedestrianism' to the language, but even if this was so he was dignifying with a new title a competitive practice which, on the evidence he presents, had been active for some considerable time: the 'age of Pedestrianism' seems to have its murky beginnings as far back as the 1760s, and has since 'been brought to great perfection by spirited individuals, especially in Britain'.[2] These men (for it was an exclusively male sport) were not part of a seamy popular culture subsisting beneath polite society, though their public exploits seem to have attracted large crowds: as Thom makes clear, most of the proponents of pedestrianism had very respectable backgrounds, and furthermore they often enjoyed the patronage of men of rank. Foster Powell, one of the most celebrated, was a lawyer by profession; while Captain Barclay, whose performances Thom documents in loving detail, was descended 'from an ancient and honourable family'[3] and only obtained a commission in the army when he had got bored with improving his Scottish estates. Although public betting on the pedestrians' expeditions and foot-matches was often lively, and although for some of the protagonists, like the impecunious and terminally luckless George Wilson,[4] money was a motivating factor, for most a gamble or challenge worth 100 or 200 guineas clearly did no more than add a gentlemanly spice to the event. Foster Powell, for one, never made much money from his walks, though it was said that he could have made thousands.[5]

Although the general tenor of Thom's book is of a piece with the genre of sports history to which it chiefly belongs (for example, he provides a long account of, and minutely detailed tables for, Barclay's walk of 1000 miles in 1000 successive hours on Newmarket heath in 1809), his promotion of pedestrianism is not grounded solely in his enjoyment of it in its sporting aspect. He has a strong belief that bodily exercise is important to every man, and that 'Exercise on foot is allowed to be the most natural and perfect, as it employs every part of the body, and effectually promotes the circulation of the blood through the arteries and veins.' More particularly, writing towards the end of the Napoleonic Wars, he is concerned that British troops are not prepared by adequate training for the hardships they will have to undergo: strengthening and augmenting the capacities of the body is all the more important at a time 'when the physical energies of many of our countrymen are frequently brought into action by the conflicts of war'.[6]

His comprehensive training regime covering sweating, exercise and feeding is therefore targeted especially at military men. He claims that the reputed feats of American Indians in going long journeys over mountainous country have been surpassed by modern Britons like Captain Barclay, and advances this as a reason for national pride and self-congratulation; his wish for such physical prowess to be emulated in the population at large, particularly within the armed forces, comes close to advocating pedestrianism as a means of empowering the nation at a critical time in its history. Walk for victory!

If this patriotic rationale for walking strains credibility, it nevertheless demonstrates one way in which a mundane physical activity can be put to work within apparently remote and more culturally privileged discourses. In Romantic writing, walking, for all its self-evident differences from other modes of travel, is rarely a parochialism of the body; and while we need to take due account of what flows from the material character of pedestrianism, we should recognise too that walking leads a mental and aesthetic life that is both distinct from, and continuous with, its bodily one. There is nothing more concrete than putting one foot in front of the other, but walking is also an idea, or a form of thinking (a way of connecting ideas), and as such can be put to a range of rhetorical uses that we need to consider.

In terms of viewing the phenomenon of Romantic pedestrianism with greater historical precision, Thom helps us to see that very serious walking had been undertaken by men from the middling orders of society, and even by the lesser gentry, well before the traditional beginning of the Romantic period; not only this, but far from compromising the social status of the participants, it seems to have redounded to their credit and made minor celebrities of some of them. Clearly it could not be in response to events in this sphere that semantic space was reorganised in the way described above in (roughly) the quarter-century before 1818. However, it is almost certainly the case that the kind of walking Thom documents is a special case, and that it was socially endorsed precisely to the extent that it was framed as a sporting spectacle (in the case of Captain Barclay's illustrious performance, *literally* framed by roping in an area of Newmarket heath). It is not with regard to this sort of activity, but to the new phenomenon of the pedestrian tour, and to other less ambitious forms of walking for pleasure undertaken by men (and, in a more limited way, women) of the professional and commercial classes, that public attitudes were challenged, and new cultural practices established, in the last ten to fifteen years of the eighteenth century.

This in itself is not a new insight. In a non-academic but well-researched study of literary pedestrians published in 1959, Morris Marples notes that 'few long-distance travellers went on foot in eighteenth-century England', but that 'a striking change' took place at the very end of the century with the rapid growth of recreational walking. Furthermore, he suggests, the pedestrians whose written testimony has survived probably represent the tip of the iceberg:

> there must have been many other people of no importance and of whom we know nothing, who were also attracted by the new fashion for walking, and from about 1790 could certainly be found enjoying themselves on foot along the roads, and even off the roads, in Wales, Scotland, and the Lake District, and on the Continent.... How many there were we have no means of knowing. But that they were fairly numerous is suggested by the fact that, from about 1800 onwards, it became worth while to publish guide-books expressly intended for pedestrians.[7]

My own researches confirm Marples's assessment, though I think it possible, owing to the volume of pedestrian travel writing generated, to give a more variegated account of Romantic walking (its prosaic norms as well as its creative exceptions) than his own concentration on the major figures allows. I also wish to probe more deeply the roots of the 'reversal of the popular attitude to walking', which Marples loosely attributes to the 'spirit of the age', noting *en passant* changes in attitudes to the natural world and the growth of aesthetic tourism.[8]

Anne Wallace's *Walking, Literature, and English Culture*, the first extended scholarly treatment of its subject, has put forward a challenging socio-economic explanation of this reversal, as part of its overall project of grounding the 'previously unrecognized literary mode' she calls 'peripatetic' (which she associates almost exclusively with Wordsworth) in 'material and literary conditions long obscured by traditional contempt for the practice of walking'.[9] I shall question Wallace's materialist rationale in due course, as indeed I shall take issue with aspects of her argument – especially her account of the role of walking in Wordsworth's poetry – throughout this book, while recognising her achievement in opening up an unfamiliar territory for critical debate. At this point I merely want to signal my concern that, despite her access to the information in Marples and elsewhere, she overstates the rarity of recreational and touristic walking in the last twenty years of the eighteenth century – saying, for instance, that 'In the early 1790s pedestrian tours like Wordsworth's in Europe and Joseph Budworth's in the Lakes are still isolated affairs', and that walking was 'thoroughly excluded from the usual practice of Grand Touring' until 'that crucial period around the turn of the nineteenth century'.[10] In what follows I may seem to be quibbling about both numbers and dates, but in the context of Wallace's thesis, as we shall see, it behoves one to be acutely sensitive both to the relative weight of evidence and to the microhistorical developments of which that evidence speaks.

What, then, is the best assessment one can form of pedestrian travel in the early Romantic period? There is less evidence available for Continental touring than for the domestic variety – not least because, as Wallace observes, the travel process is ordinarily suppressed in much contemporary travel writing, and it is remarkably difficult to determine in many cases just how the traveller *is* getting about; nevertheless, bearing in mind that one knows only of those travellers of whom some written record survives, one gets a picture of a significant amount of pedestrian activity. William Coxe, who travelled on the Continent in 1776, 1779, 1785 and 1786, and whose *Sketches of the Natural, Civil and Political State of Swisserland* went through numerous editions and undoubtedly helped Wordsworth plot the route of his pedestrian tour, was not exclusively a walker, but evidently adopted this mode of travel wherever he could. The distinction is underlined by his translator, Ramond, who followed Coxe in having 'travelled, *or rather rambled on foot* over the mountains, without following any road'.[11] Coxe's own voluntary immersion in physical discomfort seems genuine, and it is interesting to see him actively foregrounding the advantages of pedestrianism. This is still more marked in the following passage, which again explicitly contrasts walking with alternative transport:

> I walked slowly on, without envying my companions on horseback: for I could sit down upon an inviting spot, climb to the edge of a precipice, or trace a torrent by its sound. I descended at length into the *Rheinthal*, or Vally of the Rhine; the mountains of

Tyrol, which yielded neither in height or in cragginess to those of Appenzel, rising before me. And here I found a remarkable difference: for although the ascending and descending was a work of some labor; yet the variety of the scenes had given me spirits, and I was not sensible of the least fatigue. But in the plain, notwithstanding the scenery was still beautiful and picturesque, I saw at once the whole way stretching before me, and had no room for fresh expectations: I was not therefore displeased when I arrived at Oberried, after a walk of about twelve miles, my coat flung upon my shoulder like a peripatetic by profession.[12]

Coxe expresses here both the pedestrian's advantage of complete freedom of movement, and the inspiring effect of the combination of continual change of scene with maximum time for appreciation that characterises the mobile gaze of the pedestrian traveller. If not a peripatetic by profession, Coxe is clearly one by choice.

At about the same time as Coxe's first tour, or perhaps even earlier, Foster Powell took a break from athletic pedestrianism and record-breaking to go walking in France and Switzerland. In the mid-1780s William Bowles sought solace from a broken engagement in a walking tour on the Continent, taking in the north of England and Scotland before crossing to Ostend, progressing to Antwerp, then moving down the Rhine into Switzerland. He was accompanied for part of his tour by the future Earl of Cork and Orrery – a rare example of a member of the aristocracy forsaking his carriage and four. Bowles's *Sonnets*, which were to influence the young Wordsworth and Coleridge, were purportedly written on the tour, though they only intermittently assume the rhetorical situation of walking poems: 'Languid, and sad, and slow from day to day, / I journey on, yet pensive turn to view/...The streams, and vales, and hills, that steal away'.[13] In 1786, William Frend, a Cambridge don later to become a prominent dissenter and political reformer, was also on foot in France, Switzerland and Savoy with two old schoolfriends, writing letters back to his sister Mary that show a fledgling taste for the sublime. Frend had begun to experience religious doubts before leaving on his tour, but the freedom of the road seems to have accelerated his intellectual and religious emancipation, for on returning to Cambridge he was 'unwilling...to be brought under the discipline of a College',[14] and his politicisation and conversion to Unitarianism (the radical's religion in the Revolutionary era) followed soon afterwards. Joshua Wilkinson, an acquaintance of Wordsworth's, conflated his experiences of two Continental walking tours, in 1791 and 1793, in a book published several years later. A politically enthused traveller, Wilkinson, whose itinerary through France, Switzerland, Italy and Germany is remarkably similar to Wordsworth's, takes pleasure in being greeted by ordinary French citizens as an English 'friend of liberty'.[15]

Coxe, Powell, Bowles and Frend are all cast well into the shadow as pedestrian travellers by one of the most colourful figures of the period, John 'Walking' Stewart, best remembered through an affectionate essay by De Quincey. Born in London of Scottish parents, and educated at Harrow and Charterhouse, Stewart spent nearly twenty years in the East Indies, initially working as a writer for the East India Company, but later as a successful general in the army of Hyder Ali, the Muslim ruler of Mysore, then as Prime Minister for the Nabob of Arcot. Over this time he saved enough money to pursue his grand scheme of travelling the world – on foot – in search of moral truth, in the belief that 'the man who knows all the most important nations of the globe becomes the paragon of his species, and the acme of intellectual energy'.[16] Leaving India in 1783, he walked through Persia and Turkey on his long journey back to England. The following year Michael Kelly, an actor with the King's Theatre, met him in Vienna: 'The last little walk he had taken', Kelly notes ironically, 'was from Calais, through France, Italy, and the Tyrol, to Vienna, and in a few days he was going to extend it as far as Constantinople'; he adds that his 'pedestrian exploits were universally spoken of'.[17] His European travels probably continued through the 1780s, while in 1791, if no earlier, he was pedestrianising in America and Canada, his arrival being noted in a newspaper published in Albany, New York. Given even the little that is known of Stewart's peripatetic life – the paucity of information stemming from the fact that he wrote nothing about his travels, since his austere wish 'was to instruct and not to amuse people'[18] – it seems hardly an exaggeration that he was introduced to De Quincey in 1798/99 as 'a very eccentric man who had walked over the habitable globe'.[19]

I have tried to give just enough detail, where it is available, to illustrate the very different individuals who were already undertaking significant expeditions on foot in foreign parts – purposefully *choosing* this mode of travel – before Wordsworth made his 'mad and impracticable' pedestrian tour of the Continent in 1790.[20] The evidence surely tends to the conclusion that there were others like them who have left no trace of their activities: what we see is a climate of opinion in which men like Walking Stewart or Foster Powell could be lauded by contemporaries as a 'noted pedestrian' or the 'greatest pedestrian ever known in England' only against the background of the general emergence of pedestrianism as a cultural phenomenon. Moreover, the fact that the pedestrians I have identified are so diverse – Frend politically radical, Coxe a natural conservative; Stewart of an Enlightenment mentality, Bowles Romantic in temperament – argues more strongly for walking becoming gradually assimilated into mainstream culture in the last quarter of the eighteenth century.

This impression is confirmed when one turns to consider the rapid rise in domestic pedestrian touring, of which there is a much more substantial written record. Anne Wallace, in her discussion of the 'socio-economic content of walking' in the period, gives deserved prominence to the *Travels* of Carl Moritz, a German pastor who visited the north of England on foot in 1782. Moritz's account is indeed an insightful one, as well as being one of the more vivid contributions to what is a very imitative and formularised genre, and I shall shortly consider what it might tell us about contemporary attitudes to walking. But to claim, as Wallace does, that Moritz's tour and the one or two examples from the 1790s that she quotes were 'isolated affairs' is, I would argue, to slew the evidence. Lake District tourism was, of course, well established by the 1790s, stimulated by the writings of John Brown, Thomas Gray, William Gilpin, Thomas West and others, and by the vogue for picturesque travel which they set in train. But whereas such tours had most commonly been undertaken on horseback, the pedestrian alternative is now being seriously pursued and promoted: Adam Walker made the conventional tour of picturesque sites on foot in 1791, arguing that 'there is but one way of Travelling more pleasant than riding on horseback, and that is on foot; for then I can turn to the right and to the left'; Joseph Budworth followed him a year later, walking 240 miles in two weeks, and pushing the claims of domestic tourism in his published account on the grounds that 'a once-boasted, though now unfortunate, part of the Continent is become a scene of horror and devastation'; more modestly, Henry Skrine covers the Lake District in his *Three Successive Tours*, published in 1795, and although his principal mode of travel seems to be by carriage, he undertakes a lot of excursive walking and some testing mountain ascents.[21] Unlike Moritz, these writers do not complain of difficulties and derision encountered as pedestrian tourists.

One can fill out the picture by switching the focus to Wales, another area of growing popularity for tourism, serviced most notably by Thomas Pennant's *Tour in Wales* (1778). Here again the example of Wordsworth, whose life has been documented in the minutest detail by successive biographers and Mark Reed's two-volume *Chronology*, is perhaps unhelpfully luminous: his wanderings in Wales in 1791 in the company of Robert Jones, and his second journey to visit Jones in 1793, which took him on foot across Salisbury Plain and up the Wye Valley, are well known to students of the period because they lie behind famous poetry such as 'Tintern Abbey' and the description of the ascent of Snowdon in Book XIII of *The Prelude*. But Wordsworth was but one of a significant number of pedestrian tourists seeking the picturesque and sublime in Wales in the 1790s. Another, only slightly less celebrated visitor was Samuel Taylor Coleridge, who made a tour of over 600 miles in the summer of 1794 with a fellow Cambridge undergraduate, Joseph Hucks. Coleridge's letters offer a vigorously subjective perspective on the tour, which was also the subject of a workmanlike epistolary narrative by Hucks published the following year. Like Budworth, Hucks adverts to the war against France, fearing that presenting a travel book to the public at a time when 'the great scale of our political existence is in danger of sinking for ever' might appear frivolous, though it is assuredly this same political context that has given a new impetus to tourism within Britain. Hucks reminds his correspondent that the latter regarded the idea of a pedestrian tour as 'visionary and romantic',[22] but notes later that he and Coleridge have been joined for part of the trip by two other Cambridge students, which adds to the evidence that such tours were by this time part of the undergraduate subculture. Although these fellow Cantabs eventually accompany Coleridge and Hucks on foot, when first encountered they are travelling in greater comfort, which attracts ironic derision in Coleridge's letter to Robert Southey of 13 July:

> these rival *pedestrians*, perfect *Powells*, were vigorously pursuing their tour – in a *post chaise*! We laughed famously – their only excuse was, that Berdmore had *clapped* – or else etc.[23]

Coleridge's reference to Foster Powell is interesting, since it implies some interaction between the enclosed sporting world of the competitive walkers and other social groups: pedestrian tourists, especially perhaps novelty-seeking undergraduates, may well have had spice added to their expeditions by the thought of emulating the celebrated athletic performances of the period. As regards their reception, Hucks states that they met with some astonishment, 'sometimes exciting [the] risible muscles' of local people, and some alarm, being occasionally mistaken for Frenchmen, but in general they encounter no special difficulties or hostilities.[24]

The astonishment of the local inhabitants was probably muted due to the presence among them of other pedestrian tourists in the 1790s, of which it seems certain that only a percentage have left written trace of their journeys. The anonymous author of *The Observant Pedestrian* (1795) was one such: although the realities of travel experience are scarcely visible in this book through being systematically reprocessed through the conventions and affective repertoire of sentimental literature, there was presumably some real-life basis to his 'solitary tour from Caernarvon to London', and he uses the contrast between the sensitivity of both voluntary and involuntary pedestrians and the hauteur of the 'illustrious traveller' to give dramatic life to the social-moral argument.[25] 'A.A.', the signature of the chemist and mineralogist Arthur Aikin, serialised an account of a pedestrian tour in North Wales in the *Monthly Magazine* in 1796. Aikin also accompanied George Dyer on another such tour in Scotland in 1797.[26] In 1797 a Breton, Mandet de Penhouet, went on foot through South Wales, thus demonstrating the cosmopolitanism of 'lovers of the picturesque'; his correspondent had endeavoured to persuade him that a pedestrian journey 'was too full of difficulties, particularly to a foreigner', but such fears have, he reveals, proved groundless.[27] In the same year Richard Warner and a companion are also walking through Wales, and Warner advertises the advantages of the pedestrian's independent mode of travelling, which 'enables him to catch beauties in his walk through an Alpine country, which the incumbrance of a carriage, and even the indulgence of a horse, prevents another traveller from enjoying'.[28] In an amusingly paradoxical illustration of this point, Warner prefaces each chapter with a route map with dotted lines showing deviations from the main road. In 1798 the Reverend Bingley, accompanied at times by a fellow clergyman, follows in Warner's footsteps, and criticises the latter in his own *Tour* for being too novelistic; Bingley is a more serious mountaineer than most pedestrian tourists, and includes descriptions of several different ascents of Snowdon in an account that is more circumstantially realistic, and less of a sentimental journey, than others in the genre.[29]

I hope that the examples I have given have gone some way towards demonstrating that pedestrian touring in the later 1780s and the 1790s was not a matter of a few 'isolated affairs', but was a practice of rapidly growing popularity among the professional, educated classes, with the texts it generated being consumed and reviewed in the same way as other travel literature: compared, criticised for inaccuracies, assessed for topographical or antiquarian interest, and so on. Other tours could be cited: John Thelwall's in the south of England in 1797, James Plumptre's in the north of England and Scotland in 1799, and, from the very first years of the next century, John Bristed's and George Wilson's tours in the Scottish Highlands, and William Hutton's in North Wales. For the travellers of whom some record survives, there must be many who have left no trace. One of the best indices of the raised status of walking in the 1790s is a remark in the *Monthly Magazine*'s half-yearly retrospect of British literature for 1798, attached to a notice of Warner's *Walk through Wales*:

> We are happy to observe an increasing frequency of these pedestrian tours: *to walk*, is, beyond all comparison, the most independent and advantageous mode of travelling; Smelfungus and Mundungus may pursue their journey as they please; but it grieves one to see a man of taste at the mercy of a postilion.[30]

For the 'man of taste' to be actively recommended the pedestrian alternative indeed shows that a decisive reversal of educated attitudes has taken place, and within a relatively narrow span of years.

There is evidence to support Morris Marples's contention that university scholars and clergymen were prominent among these early pedestrians, but this is not universally the case. Neither should it be assumed, from the bare facts I have supplied so far, that all these tourists were completely at ease with their mode of travel, or that pedestrianism was an ideologically neutral activity in the 1790s. The situation is very much more complex, as I aim to show. But the first essential point to bear in mind is that in this very turbulent period of British and European history significant numbers of men in the middling orders of society, for whatever motives, are prepared to undertake journeys in life and text that would have been outside the mental and social horizons of the previous generation.

The aura of quixotry or unconventionality surrounding pedestrian touring towards the end of the eighteenth century has clearly dissolved by the 1810s, as is shown by the proliferation of written tours for all parts of Britain as well as (with the reopening of the Continent in 1815) an impressive range of foreign countries, and the appearance for the first time of specific information for pedestrians in general tourist guidebooks. Robert Newell's 1821 tour advertises as part of its title its *Instructions to Pedestrian Tourists*: in contrast to the titles of earlier tours, which sometimes defer mention of the mode of travel to the inside pages and sometimes offer themselves more as a curiosity or challenge to the reader by including it, this seems to assume the certainty of a market for such advice among fellow-travellers. Newell's recommendation of walking is also interesting:

> The best way undoubtedly of seeing a country is on foot. It is the safest, and most suited to every variety of road; it will often enable you to take a shorter track, and visit scenes (the finest perhaps) not otherwise accessible; it is healthy, and, with a little practice, easy; it is economical: a pedestrian is content with almost any accommodations; he, of all travellers, wants but little, 'Nor wants that little long'. And last, though not least, it is perfectly independent.[31]

Newell cites independence, as do a number of the 'first generation' of Romantic walkers I have already surveyed; more striking are his commendation of walking as the *safest* option, which reflects a very altered perception of the security of travel from that which prevailed in the eighteenth century,[32] and his advocacy of the practical and health benefits of pedestrianism, which again suggests its institutionalisation as a form of tourism and its extension to lower reaches of the middle classes.

The authors of pedestrian tours in the later Romantic period who desist altogether from defending or praising their mode of travel presumably give unwitting testimony to this process of incorporation and normalisation. In a period which sees not only the re-emergence of walking tours in the more familiar areas of

Continental Europe, but also the penetration of walkers into such inaccessible regions as Chaldaea and Siberia,[33] it is not surprising that the anonymous author of *A Tour on the Banks of the Thames from London to Oxford* should fear the reception accorded to the publication of an account of such a modest piece of pedestrianism.[34] The appearance of William Hazlitt's 'On Going a Journey', one of the masterpieces of walking literature, in *Table-Talk* in 1821, perhaps best marks the maturity of respectable pedestrianism: it is now something to be gently analysed, theorised, philosophised; not defined and defended before a suspicious or bewildered public, but fondly reflected upon for a readership hospitable to attempts to anatomise its cultivated pleasures.

Having surveyed the evolution of the walking tour in the Romantic period, insofar as this can be gauged from the published record, let us return to consider the range of contexts that help to clarify the motives and mentalities of the first wave of walkers, in the transitional period of the 1780s/1790s when inherited negative attitudes towards walking still have sufficient presence to require confronting or contesting.

In terms of the history of travel, the most immediate and obvious way in which the first generation of walkers defines itself is in contradistinction to the traditions, logistics and ethos of the Grand Tour, a phrase usually understood to apply to a circuit of France, Switzerland, Italy, Germany and the Low Countries undertaken by young gentlemen of the landed classes as part of their civilising education and preparation for public life. Some historians trace its origins as far back as the break with Rome in 1534, 'which changed spiritual pilgrims into secular tourists',[35] but for many it assumes its distinctive shape in the eighteenth century – when, at its height, an estimated 15 000–20 000 English tourists may have been abroad each year. Many young men stayed abroad for several years to complete their education, taking up residence in the major cities, so the Tour in its aristocratic guise should not be thought of as implying continuous movement from place to place. When they travelled, they did so mainly by road, and they travelled in relative comfort, either bringing their own carriages to the Continent or hiring them for the whole or part of the Tour. They tended to move from one major town or city to another along tightly-prescribed routes, with their capacity to deviate from the main roads curtailed not only by the restrictions of the posting system and the condition of road surfaces, and by other material factors such as the availability of accommodation and credit facilities, but also by the philosophy of the Tour itself. This is well summarised in the preface to Thomas Nugent's compendious guidebook, which announces the intention

> to contribute to the improvement of that noble and ancient custom of travelling, a custom so visibly tending to enrich the mind with knowledge, to rectify the judgement, to remove the prejudices of education, to compose the outward manners, and in a word to form the complete gentleman.[36]

When the goal of travelling was to improve the mind, manners and morals of the traveller – even if this 'official' rationale was at variance with the actual attractions of the Tour for many young men, such as the famously dissipated Italian lifestyle – importance would inevitably accrue to those centres of population which offered a full range of cultural and social activities; special significance attached to Rome and Florence as repositories of the classical heritage, and these cities were geared up to those seeking an artistic education.

It is a long-held view that the Grand Tour was halted in its tracks by, and never recovered from, the French Revolution – or, to be more specific, the Revolution from that stage in mid- to late 1792 when public opinion in Britain had begun to harden against it on account of its perceived descent into Terror. Most wars on the Continent earlier in the century had had little effect on tourism, because of the small-scale nature of the conflicts, but, as Jeremy Black has well described, the Revolution brought a sea-change:

> As French armies spread across the Continent, defeating Britain's allies, and new-modelling states, the Continent became an alien entity. Contacts were executed or forced to flee, British diplomatic representation withdrawn..., artistic treasures were seized by the French and old activities such as visiting nunneries, attending academies, being presented at court, and watching the ceremonies of court and religious society, ceased. The French Revolutionary government had a different attitude to British tourists in wartime than that of its *ancien régime* predecessor. Much of the Continent became less accessible, less comprehensible, and hostile, and the old-fashioned Grand Tour was a victim of this change. Tourism continued, but it followed a different course.[37]

The course it followed was one that gave a powerful new impetus to tourism within Britain, and specifically to scenic or aesthetic tourism under the aegis of the picturesque cult that I shall discuss in the next chapter. The rise of pedestrian tourism has to be situated with regard to these historical and cultural developments, though the relationship is not a straightforward one.

There is, of course, a striking difference between young aristocrats encamped in Paris or Rome and mixing in the best social circles, and whose most arduous travel experience was crossing an Alpine pass on the back of a mule, and impecunious undergraduates and lesser clergy walking anything up to 30 miles a day in search of the natural beauties of their native land. But if there is an inevitable class dimension to pedestrianism, it is not simply a matter of a middle-class revolt against the costly and well-cosseted migrations of the Grand Tourists, for the Grand Tour itself had been transformed by a progressive 'downclassing' in the course of the eighteenth century. Jeremy Black summarises it thus:

> In the first half of the century the perception of tourism was dominated by the classical Grand Tour – young men travelling with tutors for several years to Paris and Italy in order to finish their education. In the second half of the century, many still travelled in this manner, but there were also larger numbers of other tourists: travellers not on their first trip, women, older tourists, families, those of the 'middling sort' who tended to make short visits. These groups, usually unaccompanied by any guide,... did not stress education as the prime motive for travel. Instead enjoyment and amusement came increasingly to the fore.[38]

The middle classes, of course, had more limited time and money, and this determined the dramatic changes that took place in the statistical profile of the Grand Tourist: John Towner calculates, on the basis of his sample, that between the mid-sixteenth century and the 1830s the average age of the Tourist rose from 23 to 42, while the average duration of the Tour declined from 40 months to only 4 months.[39] This reduction was achieved by shortening the period of stay in particular places, rather than by faster travelling, and the shrunken Grand Tour was well established before the railways had a dramatic effect on journey times in Europe.

Enacting a dynamic which has been replayed constantly ever since, when the middle classes began to 'take over' the traditional circuit, which they did around the 1780s, the landed classes decamped in search of more exclusive destinations, such as Greece and the Near East. But my point is that pedestrian touring was an *option within* the more pervasive growth of travel for health, recreational and aesthetic purposes among the professional classes. I have been speaking of the Continental Grand Tour, but the same is manifestly true of domestic touring, for which travel on horseback or by wheeled vehicle remain options, almost certainly the dominant options, throughout the Romantic period. There is a sense in which domestic touring enters the long-running debate on the benefits of the Grand Tour, since one of the main defences of the Tour, that it improved the mind by alienating the traveller from the prejudices of upbringing and the customs and values of his native culture, was inherently weak given that most Tourists had not travelled beyond a narrow radius of home in their own country. However, we need to bear in mind that the growth of domestic touring is bound up with a number of very complex developments in the eighteenth century, including improved communications and increasing inter-regional trade, the drive to open up the country to the spirit of rational enquiry, the transformation and extension of the cultivated landscape through the enclosure movement and the accompanying change in attitudes to 'wild' nature, and so on. The more restricted observation I am making here is that, within a longer history in which, as Thomas West writes at the beginning of his *Guide to the Lakes* (1778), 'persons of genius, taste, and observation began to make the tour of their own country',[40] there arises towards the end of the eighteenth century a distinctive subculture of *pedestrian* touring. One has to take the full measure of the *voluntary* aspect of this activity: that it was a choice that perforce excluded other, invariably safer choices, and as such brought the traveller into a complex arena of social significations, and involved him/her in instabilities of identity that could be experienced either as anxiety or as liberation.

The grounds of the choices these travellers made have still to be specified adequately. Walkers were not Grand Tourists, certainly, but as a defining statement this does not go very far. If walking is something to be appraised at a symbolic level, we still need to establish the material realities with regard to which it is empowered for a period of time with this surplus symbolic value. At this point it would be helpful, I think, to return to Anne Wallace's *Walking, Literature, and English Culture* and consider the socio-economic explanation she puts forward for the appearance of 'peripatetic' as a literary genre.

Wallace's central argument is that walking could not be reconfigured and manipulated aesthetically until it was possible to think about it as something that was done by choice and at leisure, and that in turn could not come about until its strong identification with the necessitous lives of the lower orders had been dismantled, and its social connotations of poverty, unrespectability and possible criminality had started to be erased. The so-called 'transport revolution' (a phrase Wallace borrows from the historian Philip Bagwell), the first material effects of which are dated to the first decade of the nineteenth century, cleared the ground for walking to be redefined as voluntary travel, and for it to become the model for new aesthetic strategies and preferences. However, I shall let Wallace put the case herself, at what I think is necessary length:

> As travel in general becomes physically easier, faster, and less expensive, more people want and are able to arrive at more destinations with less unpleasant awareness of their travel process. At the same time the availability of an increasing range of options in conveyance, speed, price, and so forth actually encouraged comparisons of these different modes...and so an increasingly positive awareness of process that even permitted semi-nostalgic glances back at the bad old days....Then, too, although local insularity was more and more threatened...people also quite literally became more accustomed to travel and travellers, less fearful of 'foreign' ways, so that they gradually became able to regard travel as an acceptable recreation. Finally, as speeds increased and costs decreased, it simply ceased to be true that the mass of people were confined to that circle of a day's walk: they could afford both the time and the money to travel by various means and for purely recreational purposes ....And as walking became a matter of choice, it became a possible positive choice: since the common person need not travel by walking, so walking travellers need not necessarily be poor. Thus, as awareness of process became regarded as advantageous, 'economic necessity' became only one possible reading (although still sometimes a correct one) in a field of peripatetic meanings that included 'aesthetic choice'.[41]

It sounds a persuasive case. It is certainly possible that something like the shift in consciousness that Wallace describes may have taken place by the 'end' (as conventionally conceived) of the Romantic period, and influenced the spread of pedestrianism in the 1820s and 1830s; even more likely that such a shift was instrumental in shaping the attitudes of Victorians writing in the railway age, and helped generate the apostolic fervour with which writers like Leslie Stephen and Robert Louis Stevenson treated the walking tour. But it fails to account for the rise of pedestrianism as I have narrated it.

I have been at pains to demonstrate that the evidence points to walking for recreational and aesthetic purposes becoming well established during the 1790s. The process that Wallace summarises in the paragraph above must, by its very nature, have been a slow one, and the unfortunate truth is that the middle classes were walking as a matter of 'aesthetic choice' long before the altered ideological landscape that she describes may be assumed to have taken shape. Moreover, her own reading of the historical sources suggests that the *first* 'palpable effects' of the transport revolution were not evident until 'the early 1800s', and these can only be advanced as the determining factor in freeing people 'to practise and think about voluntary walking'[42] by means of the accompanying gesture in which Wallace dismisses a few earlier pedestrian tours as 'isolated affairs'. There is thus a narrow but crucial disjunction in her argument, which obscures the fact that all the pedestrians I have talked about must somehow or other have been *confronting* the still dominant prejudicial social assumptions about walking when they undertook their tours.

In case this appears too hasty a rebuttal of what is, on the surface, an attractive argument, it might be wise briefly to review the evidence, using some of the same standard secondary sources as Wallace as well as drawing upon my own reading in topographical literature of the period. The main factors in the improvement of travel conditions were the condition of the roads and vehicular technology. As regards the former, it is certainly the case that roads were improving, albeit in a very piecemeal way, throughout the second half of the eighteenth century, as a result of both administrative changes and, towards the end of the century, the revolution in road engineering pioneered by men like John Metcalfe, John MacAdam and Thomas Telford. Turnpike trusts, which make their appearance late in the seventeenth century, multiply spectacularly in the period 1750–1815. In place of a haphazard parish-based system of road repairs employing statute labour that had lasted since 1555, they placed the burden of repairs on road-users, and, in the view of Dyos and Aldcroft's survey of *British Transport*, 'made it possible for the roads to be equipped to handle the traffic which was beginning to come on to them'.[43] The improvements seem to have permeated most parts of the country: Thomas West, in the 1770s, approves the opening, 'at private expence', of new 'carriage roads for the ease and safety of such as visit the country', and states that 'the public roads are equally properly attended to'; at the turn of the century Hutton notes a great improvement in Welsh roads over the last half-century, facilitating a 'vast influx' of annual visitors.[44] Admittedly, as Dyos and Aldcroft demonstrate, opinions on roads among contemporaries are often contradictory and hard to interpret reliably, but their conclusion is that 'comparisons of road conditions prevailing in almost any part of the country at, say, twenty-year intervals after 1750 would undoubtedly show tangible improvements, expressed in terms of less intermittent use, shorter travelling times, more substantial traffic, and more generally felt effects of widespread communications'.[45] An important element in the 'more substantial traffic' is passenger traffic: Philip Bagwell estimates that between 1790 and 1836 the number of stage-coach services increased eight-fold, before taking into account enlarged vehicle capacity.[46] However, only by about 1820, in his view, was it quicker to travel by coach than on horseback, which coincides with the opinion of another transport historian that 'the period 1820–1836 can be regarded as "the golden age of coaching"'.[47]

The impression of continuous progress conveyed by a study of road improvements needs to be counterbalanced by other considerations. Firstly, as all the historians I have consulted make plain, and as is evident in travel writing of the time, faster journey times did not immediately translate into greater comfort for travellers. Eighteenth-century carriage design was such that a health warning on the door might well have been appropriate, as Bagwell confirms:

> Because the carriage body was suspended by leather straps from heavy 'perches' linked to the front and rear axles, the four inside passengers were subjected at times to a severe pendulum-like jolting. In the 1790s to be 'coached' meant getting used to the

nausea, akin to sea-sickness, which travelling in these vehicles induced.[48]

The single most important innovation in vehicular design – Bagwell compares it to the contribution of Goldsworthy Gurney's water-tube boiler in the 1820s to the age of steam – was Obadiah Elliot's elliptical spring, patented in 1805, which made possible lighter, speedier, and more comfortable carriages. As Elliot's invention was taken up, allowing for increased passenger traffic (safer carriages meant that more outside passengers could be legally accommodated) and paving the way for the 'golden age' of coaching, it might have begun to alter perception of the travel process in the way Wallace suggests, but given the chronological data such shifts in perception are not imaginable before the latter part of the Romantic period.

The second qualification one needs to enter concerns the accessibility of improved travel to the community at large. Throughout the period, stage-coaches and mail coaches were still the preserve of a small social élite ('travelling post', either on horseback or by post-chaise, was an even more expensive option). Dyos and Aldcroft state bluntly that 'Travelling by road was...within the reach of a tiny fraction only of the total population, as a comparison of the seating capacity available to the inhabitants of any sizeable town shows.'[49] Stage-coaching cost about $2d$ or $3d$ per mile; the silk-mill workers whom John Thelwall met on his pedestrian tour in 1797 earned no more than $10s$ a week.[50] Clearly even a return journey to a town a few miles away would have been inconceivable to them. The large, slow stage-waggons could be as cheap as $\frac{1}{2}d$ per mile, but they were primarily goods vehicles and passengers were often expected to walk alongside them. Of course, the labouring classes in this period did not often have occasion to go outside their parish, but this merely underlines the fact that long-distance travelling by anything other than pedestrian means was a socially exclusive activity. When the lower orders travelled, they went on foot, as they had always done; and this meant that, until the popularisation of middle-class pedestrianism, walking was an almost unmistakable index of poverty.

One final aspect of the 'transport revolution' is worth mentioning. A stage-waggon ground along at no more than 2 mph. A stage-coach in the early Romantic period might manage 4 mph; one of the new mail-coaches might get up to 7 mph on a good surface. Even in the 1820s, with the new-model carriages and macadamised roads, a coach would be doing well to average over 10 mph. I would not wish to underestimate the impact on the mind of the Romantic traveller of such variations in speed, however negligible they might be to modern senses anaesthetised by motorway driving: Thelwall, after a bad lunch, could not face continuing his walk and took the coach from Whitchurch to Salisbury in 1797; he felt that he was 'flying' to his destination, and that places along the way 'passed like so many meteors'.[51] Nevertheless, it is worth pointing out that, at a speed of 4 mph, a stage-coach would have been overtaken by Foster Powell or Barclay on one of their pedestrian expeditions. Leigh Hunt, writing around 1830, tells an amusing story of a man who was riding downhill in a post-chaise when the bottom fell out, and bystanders 'observed a couple of legs underneath, emulating, with all their might, the rapidity of the wheels'.[52] Given the average speed of a post-chaise, even downhill, this Flintstones-like scenario is not as improbable as it sounds. In terms, too, of the mapping of time on to space, the distinction between pedestrian touring and other modes of travel is a difficult one to grasp in absolute terms: an average day's work for stage-horses (not using the posting system) in the 1790s was 25 miles, whereas Wordsworth regularly exceeded this on his Continental walking tour.

It seems likely, therefore, that the difference between walking and passenger travel in the Romantic period – *a fortiori* in the early part – is not reliably articulated in terms of such factors as speed and comfort. Differences were certainly *felt*, as they were between vehicular journeys at opposite ends of the period, with what seem to us marginal improvements in speed, but it may be that these are best approached via a consideration of the complete sensuous and perceptual experience of different modes of travel. What I want to insist on here is that, far from 'easier, faster, and less expensive' travel lying behind the change in attitudes to walking, as Wallace contends, the very reverse can be argued. In the 1790s, when the pedestrian tour comes into fashion, passenger travel is slow, uncomfortable, and limited to a tiny minority of the population. The choice of walking rather than any other means cannot be explained by any crudely materialist calculus; instead one has to concentrate on what walking *signified*.

This may well seem perplexing. For what it signified, at the simplest level, as I have already made clear, was poverty, unrespectability and possible criminal intent. Conversely, passenger travel by any but the cheapest vehicular means connoted money and status. The first generation of pedestrians (a term I am here applying to those in the educated classes who walked out of choice) had to confront these associations: it was possible to exploit them, as I shall argue, or to defend against them, or simply to present a bold front, but none could have been unaware of them. Throughout history, the travels of the poor have been involuntary or customary, a story of forced migrations or the perpetual displacements of nonterritorial communities. Little has changed: Eric Leed reports a survey of Americans abroad in 1986 which merely proves that '"travel for pleasure" remains as it was through the ages, a mark of success and status, while travel under compulsion of some necessity or in service of need is a mark of commonality and a common fate'.[53] Such involuntary, depersonalising travel has invariably taken place on foot. (As I write this chapter, in 1994, a million Rwandan refugees have walked across the border into neighbouring Zaire.) Within the specifically British context, the right to *voluntary* departure is established as a mark of legal freedom in medieval times: 'According to the laws of Henry II', Leed writes, 'a lord who wished to free his serf had first to declare that intention in a church, a market, or a county court, to bestow a sword or a lance upon his former bondsman, and then take him to a crossroads to show him that "all ways lie open to his feet".'[54] The association of walking with freedom here is rare before the Romantic period; but even if it is not a purely symbolic expression, the concatenation was not to last. For a new twist was given to this history in the later Middle Ages with the introduction of stringent travel restrictions aimed at controlling the movement of labour. There were a number of permitted exceptions to these laws, but any traveller not transparently of genteel status could come under suspicion; and, as Wallace points out, 'The greatest suspicion of all fell on pedestrians, whose mode of travel proclaimed their poverty and therefore the greater probability of their being wanderers with some illicit or economically disruptive motive.'[55] The word 'footpad' is an obvious linguistic trace of the criminal taint that attached to walking: in contrast to highwaymen, who acquired a falsely romantic reputation, the footpad's horseless predations placed him at the bottom of the social scale.

It is against this deeply-sedimented history of associating walking with indigence, necessity and fate, and in later centuries with the illicit freedom of the road and the deterrent force of the still-active vagrancy laws, that the early Romantic pedestrians set out on their tours and expeditions. The most vivid and disquieting account of what this might entail is that of Pastor Moritz, who made his way on foot to the north of England in 1782. Seemingly completely unprepared by Continental circumstances for what he will encounter, Moritz is having doubts no further from London than Windsor, finding a pedestrian 'considered as a sort of wild man, or an out-of-the-way-being, who is stared at, pitied, suspected, and shunned by every body that meets him'.[56] To his disgust, he is unable to read Milton by the side of the road without being stared at by those who ride or drive by. He bemoans the trials of pedestrians in a 'land of carriages and horses', these trials including a footpad who extorts a shilling from him. He tries unsuccessfully to find a bed for the night in Nuneham and has to walk a further five miles to Oxford, but is fortunate in falling in with a clergyman who takes him to stay at The Mitre, where he spends the night carousing with other clerics. Next morning the landlady admires his courage in travelling on foot, and his motives (to learn more of men and manners in England), but admits that he would not have been given a bed if he had not been introduced as he was:

> I was now confirmed in my suspicions, that, in England, any person undertaking so long a journey on foot, is sure to be looked upon, and considered as either a beggar, or a vagabond, or some necessitous wretch, which is a character not much more popular than that of a rogue.... But, with all my partiality for this country, it is impossible, even in theory, and much less so in practice, to approve of a system which confines all the pleasures and benefits of travel to the rich. A poor peripatetic is hardly allowed even the humble merit of being honest.[57]

A fellow-traveller, in reply to the question of why Englishmen do not occasionally, 'merely to see life in every point of view, travel on foot', replies that they are 'too rich, too lazy, and too proud'. Moritz muses in agreement that 'the poorest Englishman one sees, is prouder and better pleased to expose himself to the danger of having his neck broken, on the outside of a stage, than to walk any considerable distance, though he might walk ever so much at his ease'.[58] Although the 'poorest Englishman' Moritz saw was unlikely to have sufficient money to travel by stage-coach, he would know what this mode of travel would confer on him if he had: respectability and status.

Even if there is no rhetorical inflation in Moritz's account of the suspicion and hostility he incurs, it is impossible to know to what extent his position as a foreigner accounts for his difficulties. However, there is evidence from other travellers to support his picture of the cultural minefield awaiting the pedestrian tourist in the 1780s and 1790s. Joseph Hucks, Coleridge's companion in North Wales in 1794, gives the impression that their survival strategy is one of protective colouration: rather than advertise the discrepancy between their rank and their mode of travel, 'ideas of appearance and gentility' are said to be 'entirely out of the question', and this 'metamorphosis' (romantically caricatured) has the advantage of allowing them 'to see, not to be seen'.

> and if I thought I had one acquaintance who would be ashamed of me and my knapsack, seated by the fire side of an honest Welsh peasant, in a country village, I should not only make myself perfectly easy on my own account, but should be induced to pity and despise him for his weakness.[59]

It seems that it was easier to avoid attention and opprobrium in this way in Wales than it was on the more densely-travelled and class-conscious routes in England that Moritz followed. De Quincey, reminiscing in his revised *Confessions* on his own walking tour of the Principality in 1802, affirms that 'no sort of disgrace attached in Wales, as too generally upon the great roads of England, to the pedestrian style of travelling'; interestingly, he adds that 'the majority' of fellow-tourists he met were travelling in this way.[60] Despite this, Richard Warner, who has also strategically 'dressed down', was anxious about looking like a pedlar and experienced feelings of 'false shame' when encountering 'fashionable ladies' in the course of his *Walk through Wales*, despite having worked himself into 'a fancied heroism in this respect'; he is driven to make excuses for choosing the pedestrian mode.[61]

It is probable that English tourists were unsure what to expect in Wales; but the evidence is that their sensibilities were conditioned by what they knew of the hostile climate surrounding pedestrians in their own part of the country, and their defence mechanisms were set up accordingly. Thelwall's comments on finding a 'decent, humble, but comfortable' inn on the outskirts of Basingstoke confirm Moritz's account of the situation in England: the inn, he says, was

> just such a one as the pedestrian may regard as a prize in the lottery. No swaggering post-boy to jostle him from the fire, no powdered waiter to sneer at his dusty garb, no pursey landlady to measure him, with her eye, from head to foot, and inquire for his horses, or his carriage![62]

There is therefore no question of the prejudice that pedestrian travellers would confront in England in the 1790s, though attitudes were to change over the next 20 years. In Wales and Scotland the situation was a little more complex, with tourists inevitably carrying their acculturated expectations and responses with them into a less inhospitable setting, but with some tourists making efforts to erase the marks of gentility, either out of self-defence or out of a desire to facilitate their pseudo-anthropological missions. John Bristed is one of those who, showing a blend of the spirit of rational enquiry with a Romantic eagerness to hunt out the natural man, tried to minimise the curiosity and wariness he believed he would arouse as a gentleman-traveller by adopting a form of disguise. Bristed's check shirt, cat-skin cap and green spectacles would not have allowed him to blend in unnoticeably, but were part of an attempt to pass himself off as an American sailor (in the belief that the Scots disliked the English and the Irish); he imagined that he and his companion were travelling in such a way 'as would, in all probability, induce the people whom we met to treat us without any disguise of factitious and artificial civility, and show their native character, whatever it may be, in all its outlines and features'.[63]

One final useful insight into the class context of pedestrianism in the period is furnished by a tour which, with the alienated eye of a true American, openly refers contrasting attitudes to objective differences in physical and economic conditions. Philip Stansbury, who undertook an impressive pedestrian tour of over 2000 miles in North America in 1821, remarks on the commonness of walking in 'foreign countries', including Scotland, 'where the cost of riding is proportionally high, money hard to be procured, and where the roads are excellent'; in America, however, where 'every man lives in abundance, and possesses at all times, the most ample means of self-transportation', pedestrianism is rare. He adds:

Here, it is seldom indeed, that we meet a solitary passenger, and when we do meet one, we believe, he travels in that manner, either through inclination and pleasure, or out of a principle of the strictest economy; and whatever may be his reasons in reality, for proceeding so, *we know not how to rate him from the circumstance*, nor can we act towards him with less civility, than if he made his appearance on the back of a horse.[64]

If Stansbury had travelled in Britain, as he seems to have done, then he had first-hand experience of a context for recreational and touristic walking very different to the one he assigns to America – where, he implies, one cannot reliably make assumptions about the rank or status of any pedestrian from his mode of travel. In Britain, and pre-eminently in England, at least in the early Romantic period, people knew, or thought they knew, exactly how to 'rate' someone who was travelling on foot.

I may seem to have compounded the problem rather than resolved it. Given the mistrust, intolerance and discrimination that awaited pedestrian travellers in the 1780s and 1790s, and the number of years it would take to destigmatise an activity later to be practised with religious fervour by the Victorian middle classes, how can one explain the *choice* of that first generation to travel in this way, involving themselves in difficulties and discomforts and endangering their social standing? What drove them to want to see their country, or foreign countries, in a way that their parents and grandparents would not have contemplated? Most of the answers to this question – and it seems probable that the choice is an overdetermined one in most individual cases – will have to be reserved for the next chapter, but it seems inescapable, to round off the discussion of class, that there was an element of deliberate social nonconformism, of oppositionality, in the self-levelling expeditions of most early pedestrians. When many of them were freethinking undergraduates, as yet unincorporated into professional value systems and economic subservience, and members of the lower clergy, a traditionally oppressed and disgruntled social group, this is perhaps not surprising; but the catchment area for pedestrian travellers was wider than this, as I have shown, and it may be legitimate to speculate on the deeper and more general motivation to independent travel which these groups express with unusual frequency and clarity. Eric Leed, who does not deal with the specific issue of pedestrian touring within his magisterial philosophical history of travel, nevertheless provides valuable conceptual guidelines in discussing how the Romantic period witnessed a dissociation of the idea of travel from the meanings integral to its parent linguistic form, 'travail':

> Travel became distinguishable from pain and began to be regarded as an intellectual pleasure.... These factors – the voluntariness of departure, the freedom implicit in the indeterminacies of mobility, the pleasure of travel free from necessity, the notion that travel signifies autonomy and is a means for demonstrating what one 'really' is independent of one context or set of defining associations – remain the characteristics of the modern conception of travel.[65]

I believe this gets as near to encapsulating the impetus of the first generation of pedestrians as any brief formula can. These men were not walking because the ideological space had been cleared for them to do so as a result of social and technological change; instead they were intent upon clearing their own ideological space. In a cultural landscape where there was strong peer and class pressure to declare one's status and income in the manner of one's travelling, their walking was a radical assertion of autonomy: they travelled in a way they did not have to, and in a way they could not be suspected (maugre the occasional inept disguises) of having to adopt from necessity. Walking affirmed a desired freedom from context, however partial, temporary or illusory that freedom might be: freedom from the context of their upbringing and education, the context of parental expectations and class etiquette, the context of a hierarchical and segregated society. Freedom, finally, from a culturally defined and circumscribed self. This promise was implicit in what Leed nicely calls the 'indeterminacies of mobility', which the pedestrian traveller enjoys to a greater extent than any other. It is a point that provides a convenient bridge to my next chapter.

2
An Anatomy of the Pedestrian Traveller

When the many pedestrian tourists introduced in Chapter 1 thought it proper in their written accounts to justify their mode of travel, their defence was invariably conducted in terms of the more complete freedom of movement permitted to the pedestrian. Adam Walker claims Rousseau's support for the belief that 'there is but one way of Travelling more pleasant than riding on horseback, and that is on foot; for then I can turn to the right and to the left'; Joseph Hucks gives the motives for his continuing preference as 'convenience and independency'; and Mandet de Penhouet identifies the advantage of the foot-traveller over the carriage passenger as lying in a greater freedom to satisfy one's curiosity along the way.[1] William Bingley, writing at the end of the 1790s, spiritedly restates the case that is now becoming almost formulaic. Pedestrianism, he claims, is the most 'useful' mode of travel, 'if health and strength are not wanting':

> To a naturalist, it is evidently so; since, by this means, he is enabled to examine the country as he goes along; and when he sees occasion, he can also strike out of the road, amongst the mountains or morasses, in a manner completely independent of all those obstacles that inevitably attend the bringing of carriages or horses.[2]

Bingley has a specific reason here for valuing the combination of freedom and intimacy with one's surroundings enjoyed by the pedestrian, but his rationale is generalisable to other travellers. In the first chapter I offered evidence that by the 1820s, when the benefits of walking in terms of health, safety and financial economy are also being confidently trumpeted, this association of pedestrianism with freedom and independence has hardened into an ideology which to all appearances has wide social acceptance. What begins as a slightly risqué emancipation of the middle-class traveller from the privileges and 'comforts' of the horse or carriage ends in flamboyant quixotry and self-caricature, as in the anonymous *Recollections of a Pedestrian* published in 1826:

> I have hitherto been a great wanderer. But I put not my trust in chariots or horses; I prefer the *pede libero*, the free unshackled enjoyment and disposal of my time and my limbs, and not to have the one at the discretion of the *Conducteur de Diligence*, nor the others cramped and tortured in stocks upon wheels.[3]

The priority accorded by these early pedestrians to independence and freedom of movement needs to be given its full weight. Certainly such values cohere with the principle of autonomy I outlined at the end of Chapter 1, and there may be more than a playful analogy to be drawn between the freedom to deviate – to 'strike out', to 'turn to the right and the left' – in one's route, on the one hand, and the kind of moderate *social* deviance or class nonconformism I have imputed to the first generation of pedestrians. Improved roads, after all, were one of the principal means by which the country was building a national communications network that would underpin the huge commercial and industrial expansion of the nineteenth century; changing the landscape of the country to produce the arterial interconnection of the modern state in place of a geography of more or less self-enclosed local communities; consolidating the administrative structures of the state and facilitating political hegemony over a rapidly growing and potentially unstable population;[4] and promulgating a 'national' culture in face of regional diversity and independence. With the main roads such powerful instruments of change, the walker's decision to exploit his freedom to resist the imperative of destination and explore instead the lanes, by-roads and field-paths, could well be interpreted as an act of denial, flight or dissent *vis-à-vis* the forces that were ineradicably transforming British society.

The pedestrian's freedom of movement is given a very different inflection in one of the more curious literary productions of the 1790s, John Thelwall's *The Peripatetic*. This is a hybrid of the sentimental novel, pseudo-autobiographical journal, and collection of essays, recounting in prose and verse the wanderings of Thelwall's alter-ego, Sylvanus Theophrastus. In setting out on a pedestrian expedition, Thelwall aligns himself with the popular tradition associating philosophy with peripatetic activity that began in the ancient world:

> In one respect, at least, said I, after quitting the public road, in order to pursue a path, faintly tracked through the luxuriant herbage of the fields, and which left me at liberty to indulge the solitary reveries of a mind, to which the volume of nature is ever open at some page of instruction and delight; – In one respect, at least, I may boast of a resemblance to the simplicity of the ancient sages: I pursue my meditations on foot, and can find occasion for philosophic reflection, wherever yon fretted vault (the philosopher's best canopy) extends its glorious covering.[5]

Here, the guiding idea is that since the proper object of a philosopher's study is the 'volume of nature', that condition is most appropriate which brings him into closest touch with that multifarious volume; but this is compounded by the more material reference to 'quitting the public road' in favour of the deserted field-paths, which implies that the intellectual freedom that will characterise Thelwall's writing is possible only via an initial wilful self-isolation from the structures of social life. Thelwall is repeating the programme that had been undertaken more recently in Rousseau's *Reveries of the Solitary Walker*, where the author's postscript to his *Confessions* takes the shape of a record of his 'solitary walks', these being the only time 'when I give free rein to my thoughts and let my ideas follow their natural course, unrestricted and unconfined'.[6]

Later in his expedition (having been joined by the liberal-minded 'Ambulator'), Thelwall excuses a lengthy digression on his early years, his literary apprenticeship, and the need for confessional frankness, by alluding to the pedestrian structure of the work:

> If...I should appear, according to thy better judgment, to have wandered too far from the point, thou wilt be kind enough to remember, that, as I am only a foot traveller, the bye path to the right and to the left is always as open to me as the turnpike road: and that if, on the present occasion, I have been rambling somewhat too long among the fields and green allies of poetical digression, thou art, nevertheless, bound in gratitude to excuse me, since I have been induced so to do purely for thy sake, and

to give thee to understand...what sort of a fellow he is who professes to entertain thee.[7]

The repeated representation of textual progress as a departure from the main road into solitary rambling along paths and lanes is here negatively marked as 'digression', but this is swiftly recuperated, in a manner characteristic of Romantic writing from Rousseau onwards, by incorporating the digression into the central project of the work.

In these two passages from *The Peripatetic*, one sees a direct equation between free philosophical reflection and an experimental literary style, on the one hand, and a particular kind of pedestrian practice, on the other, and it seems misguided to view the latter as just loosely figurative of the former. I make this point because I again find myself in disagreement with Anne Wallace, who admires Thelwall's book in so far as it displays emergent features of the peripatetic aesthetic she espouses, but finds it disappointingly void of attention to the details of travel process. For her, *The Peripatetic* is another example of the broad tendency to which Rousseau's *Reveries* belong, in which ' "walking" has been dephysicalized into analogy and its material components separated from the intellectual and spiritual growth it "stands for" '.[8]

To my mind, Wallace's near-obsession with the subordination of passage to destination in travel writing, and the accompanying valorisation of literature which simulates pedestrian perspectives, leads her to impoverish our appreciation of the complex relation of walking to thought and writing. In fact, she wants to prioritise a *particular* dephysicalisation of walking into aesthetic form (the neogeorgic equation of walking with cultivating labour that is the heart of her reading of Wordsworth), but this brings with it a needless devaluing of other possibilities. Intentionally or not, she presents body and mind as an absolute dualism, and in so doing underestimates the fascinating congruence between bodily action and mental process which Thelwall and many other Romantic writers articulate. I prefer the approach taken by Eric Leed in his thought-provoking analysis of the traveller's 'experience of passage', in particular the effects of motion in hardening the outlines of subject and object:

> The mind of the traveller is not separate from the body of the traveller.... The mental effects of passage – the development of

An Anatomy of the Pedestrian Traveller 33

observational skills, the concentration on forms and relations, the sense of distance between an observing self and a world of objects perceived first in their materiality, their externalities and surfaces, the subjectivity of the observer – are inseparable from the physical conditions of movement through space.[9]

Leed's remarks are addressed to the traveller in general, not specifically the pedestrian, but the philosophical attitude here is sympathetic to my reading of the ways in which intellectual processes and textual effects are grounded in the material practice of walking: the 'physical conditions' of the pedestrian's 'movement through space' are decidedly different from those of other travellers, but they too yield a range of mentalities and a variety of aesthetic forms. This is not to imply some organic oneness of sense and expression in peripatetic literature, but to insist that in the displacement from physical experience to the order of imagined reality and literary representation the rhythms and modalities of walking remain a visibly determining influence.

The freedom of movement and independence constantly adduced as motives for pedestrian travel therefore need to be taken seriously, and one should not underrate their importance as the bodily manifestation of social or intellectual energies discharged also in more conventional ways. But this is to provide only a partial anatomy of the pedestrian traveller; there are several other elements in the loose assembly of motives and desires constituting that historical figure that need to be considered. However, the next step in this analysis will entail resuming my discussion of John Thelwall.

RADICAL WALKING

Jeffrey Robinson, one of the handful of critics to have taken note of Thelwall's *Peripatetic*, makes the penetrating observation that in this text 'walking signifies the restlessness, negatively the uprootedness and political drivenness, more positively the *mobility*, of the radical mind'.[10] This rather intensifies my speculations above on the possible anticonformist dimension to the pedestrian's pleasure in a devious route and deferral of destination. It suggests that one form of response to the negative class associations of walking I described in Chapter 1 was to aggressively harness these social significations in the service of radical politics: a kind of extra-parliamentary direct action that might be expressed purely as the subversive theatricality of an educated gentleman 'levelling' himself with the poor in his mode of travel, or as an earnest desire to break down the barriers of mutual ignorance and suspicion between classes by enquiring into the lives of the lower orders in a more careful and intimate way than would be possible by using 'respectable' means of transport.

There is ample evidence of the 'political drivenness' of much walking in the 1780s and especially the 1790s, whether in the (at first sight) chance coalition of pedestrianism with political radicalism in certain individuals or in more expressive cases like those of Thelwall and Wordsworth. I have already referred to the Continental walking tour undertaken by the dissenter William Frend, whose 1793 pamphlet, *Peace and Union*, attacking the British declaration of war on France as well as the government's fiscal policy, led to a show trial at which he was expelled from his Cambridge fellowship. Another, less dramatic case might be that of the eccentric clergyman and poet William Crowe, author of *Lewesdon Hill*, who was said by Tom Moore to be 'ultra-whig, almost a republican' in politics, and who was also an inveterate pedestrian – for years regularly walking from his living in Wiltshire to Oxford, where he held the post of public orator.[11] These are men whose 'mobile' political sympathies might well be seen as finding satisfactory displaced expression in the rigours of pedestrian motion.

Coleridge and Hucks present a better-documented case of radical walking. Their walking tour took them from Cambridge to North Wales via Oxford, where Coleridge first hatched with Robert Southey the plan for an emigrant communitarian Utopia he called Pantisocracy. Though Hucks was not to be one of the original Pantisocrats, it is likely that he was caught up in Coleridge's visionary fervour, which undoubtedly sharpened the political edge of the whole tour. I shall be treating Coleridge at length in Chapter 5, so here I shall simply note the conjunction in his letters written on the tour between the idea of walking as social levelling and the idea of Pantisocracy: his first letter to Southey includes a draft of 'Perspiration, a Travelling Eclogue', in which he draws a harsh distinction between the physical exposure of the pedestrian ('scorching to th'unwary Traveller's touch / The stone-fence flings it's narrow Slip of Shade') and the 'clatt'ring Wheels' of 'Loath'd Aristocracy';

while he later reports that he has 'done nothing but dream of the System of no Property every step of the Way'.[12] Meanwhile, among the many (not easily compatible) strands in Hucks's *Tour* is a sincere and indignant radicalism of his own, which leads him to denounce poverty and hereditary privilege and the abrogation of the 'social compact' by 'vicious and corrupted governments', and to inveigh against the 'calamitous and destructive war' against France with its accompanying erosion of civil liberties at home. More to the point, Hucks clearly saw an important part of his purpose on the pedestrian tour as *learning* first-hand about the conditions of life among the lower orders in Wales, and the outcome of this self-directed consciousness-raising is a more pungent critique of the effects of poverty in terms of ill-health and early mortality, and an assault on the government which 'ought to remove every obstacle to population', but instead 'endeavours to depress and retard it'.[13]

As I noted in Chapter 1, Coleridge and Hucks were joined for part of the tour by two other Cambridge undergraduates, and in the highly politicised atmosphere at Cambridge in the early years of the French Revolution it is likely that radical walking of this kind was something of a vogue. Wordsworth's 1790 walk through France and Switzerland is another case in point, though he played down political motives in the account of the tour written some fourteen years later for *The Prelude*. I shall examine the literary output of that tour in detail in Chapter 4, so here it is better to give a little more attention to the non-university-educated John Thelwall, whose pedestrianism was cradled in a fiercely practical political engagement more long-lasting than that of any of his contemporaries in the revolutionary decade.[14]

By the time Thelwall undertook his walking tour of England and Wales in the summer of 1797 he had retired from public life, battered into submission, along with most of the radical movement, by the unremitting attentions of the state; but the narrative of his travels in the *Monthly Magazine* leaves no doubt that they were fuelled by a political passion little diminished in its intensity. As with many early pedestrians, Thelwall's is an overdetermined journey, with antiquarian and picturesque interests superadded to a desire to make personal contact with Coleridge, with whom he had corresponded for some time; but the preface to his journal gives due weight to the political information-gathering motive that was central to radical walking:

Every fact connected with the history and actual condition of the laborious classes had become important to a heart throbbing with anxiety for the welfare of the human race: and facts of this description are not to be collected by remaining, 'like a homely weed, fixed to one spot'.[15]

The son of a silk-mercer, and himself a tailor-turned-journalist, Thelwall was a metropolitan figure who had an understandable desire to familiarise himself with lower-class life in the more rural areas. The overriding political theme of his Tour is consequently the destructive impact of agrarian monopoly on the lives of small farmers and peasant labourers. His observations at Amesbury in Wiltshire are typically blunt:

There are three or four individuals in this neighbourhood, who rent to the amount of 1000l. a year each: that is to say, so many agricultural canibals, who have devoured their eight or ten families a piece.[16]

The silk-mill he comes across at Overton reinforces his conviction of 'the evil of the manufacturing system': his enquiries elicit details of the low wages, including the reduced rates paid to a large number of children aged between five and fifteen, and he reports his fears that those children who 'survive the contagion of their prison-house' turn of necessity to prostitution or soldiering when they reach adulthood. These kinds of observation and reflection are the regular and natural outcrop of his 'politically driven' tour, which provides for more patient and careful investigation than if he were travelling by coach.

The content of Thelwall's narrative is far from homogeneous, however. Although it is clear that, for Thelwall, walking loosens and stimulates the mind, the intellectual mobility thus achieved is not bounded by political subjects. Early on in his tour he writes of a 'vivacious' conversation covering 'various subjects of literature and criticism, the state of morals and the existing institutions of society' that occupied a morning's walk, and had the reciprocal effect of making 'the miles pass unheeded under our feet'. Paradoxically, locomotion in this instance generates and matures 'Utopian plans of retirement and colonisations', of 'spending the remainder of our days in rustic industry and philosophical seclusion'.[17] These ideas had only 'floated across' the brain before, in more sedentary moments of life: it is a fact attested by several writers on walking that the instabilities of motion can nevertheless make possible an intense and concentrated inwardness; here the irony is compounded by the direction of this focused reflection to thoughts of a static, rooted, contemplative life. This is a local instance of the dialectical relation between mobility and emplacement that orders a good deal of travel writing, as well as a fair amount of other prose and poetry addressing pedestrian themes that I shall be looking at: it is a matter of an ongoing, irresolvable tension between contrary human longings, well phrased by Eric Leed as those of 'motion and rest, liberty and confinement, indeterminacy and definition'.[18] Would the satisfactions of life in a small, self-sufficient pastoral community in Pennsylvania such as the Pantisocrats fantasised have appeased the migratory urge that took them there in the first place?

Mobility of mind is also expressed in Thelwall's tour by the many bizarre transitions and juxtapositions that occur without comment. At Salisbury, for example, his aesthetic contemplation of the stained-glass windows in the cathedral goes side by side with another exposure of the evils of child labour in the wool factories at Quidhampton. Throughout there is a similar sequential ordering of different forms of attention, and Thelwall either does not feel, or chooses not to advertise, the resulting incongruities. Like many Romantic tourists, his is a variegated mind, and the pedestrian tour-structure allows him to bring his various interests into play successively, without the pretence of a fully unified sensibility.

Of course, these other aspects of the mobility and prospectivism said to characterise the radical mind are not themselves intrinsically political, and so not directly relevant to the theme of radical walking; but they help to convey the uneasily constructed subjectivity of a man honest enough not to dissimulate his educated, high-cultural tastes beneath a bogusly uniform demagoguery. Since travel detaches the individual from their place in the social structure, and loosens the moorings of their culturally-constructed self, such ambiguities are perhaps more easily expressed in passage than in a settled condition. In fact, this quality of the travelling state is one of many in the profile of the Romantic walker that can be plotted against the characteristics of the 'liminal' phase of rites of passage, as analysed by the anthropologist Victor Turner. According to Turner, such rites involve three phases of 'separation', 'liminality' and 'aggregation': in the first phase, ritual subjects are symbolically isolated from their social group or from a certain cultural state; in the second phase, they enter a fluid and undefined state where customary norms and standards are not operative; in the third, their passage is 'consummated' with re-entry (at a different level) to a hierarchical social system with all the entailed rights and duties.[19] Liminal subjects are said to develop an 'intense comradeship and egalitarianism', and to bring into focus an alternative model of society as 'an unstructured or rudimentarily structured and relatively undifferentiated *comitatus*, community, or even communion of equal individuals', to set against the ruling hierarchical system;[20] the values of *communitas* they learn, and invest with a measure of sacredness, are carried over in diluted form to the aggregation phase, and find diverse manifestations in later life.

Among other oppositions that Turner sees as expressing the difference between liminality and the status system are the following: disregard for personal appearance/care for personal appearance; no distinctions of wealth/distinctions of wealth; sacredness/secularity; suspension of kinship rights and obligations/presence of kinship rights and obligations; continuous reference to mystical powers/intermittent reference to such powers; simplicity/complexity; acceptance of pain and suffering/avoidance of pain and suffering. The 'liminal' terms of these oppositions correlate remarkably well, albeit sometimes in a rather parodically secularised way, with the qualities of the radical walkers of the early Romantic period. These men made a conscious effort to 'level' themselves socially, disregarding their appearance or actively disguising their gentility, wilfully embracing the hardships of pedestrian travel, and setting off in many cases in defiance of family expectations and inherited standards of behaviour; they undertook their journeys in the name of, or in the spirit of, values – such as that of universal brotherhood – thought to be submerged in their own society; and their gravitation to the remoter parts of the kingdom answered to a Romantic-primitivist interest in the simple and unsophisticated, as well as to a yearning to experience the 'mystical powers' of sublime natural scenery. One might therefore argue that the pedestrian tour offered liberal- or radically-minded young Romantics a particularly satisfying rite of passage, and that for some the specific form of the rite, no less than its ethos, was reproduced in later life. The correlation is obviously neatest in the case of the free-thinking undergraduate, since university as a global experience has in any case traditionally been conceived of as a rite of passage, but older men like Thelwall or radical clerics like Frend and Crowe can be viewed as rediscovering *communitas* as they walked in a spirit of disenchantment with the social, economic and political order and their own roles within it.

THE PILGRIM-WALKER

> Behold us, then, more like two pilgrims performing a journey to the tomb of some wonder-working saint, than men travelling for their pleasure and amusement.[21]

Joseph Hucks writes self-mockingly of the resemblance he and Coleridge bore to pilgrims, but this identification, implied if not fully explicit as in Hucks's tour, is not uncommon among Romantic walkers and deserves consideration. If many of the characteristics of 'liminality' as outlined above seem appropriate to the quasi-ritual self-levelling and social estrangement of the radical walker, they also apply convincingly to the condition of the pilgrim, as Eric Leed's summary of the rules of pilgrimage to Palestine in the early medieval period helps to clarify:

> They were to carry no weapons and to journey barefoot, clothed with the simple, rough robe, broad-brimmed hat, and wallet that soon became their identifying costume. They were encouraged to fast, to abstain from meat, and never to spend more than one night in any one location. In addition, pilgrims were admonished to avoid iron utensils, to neglect caring for the hair and fingernails, and to shun warm baths and soft beds. The poverty of the pilgrims, besides being a holy state, also made them unprofitable prey for the lords and warbands who infested the roads.[22]

In time, of course, pilgrimage lost its ascetic purity – the pilgrims who set out from the Tabard in Chaucer's *Canterbury Tales* rode rather than walked, for example – and acquired many of the trappings of modern industrialised tourism, including resident guides at the most popular shrines and the sale of souvenirs in the form of relics. Certainly, the religious ferment of the late eighteenth century witnessed a return of back-to-basics pilgrimage on some scale: Robert Southey tells the fascinating story of John Wright, a carpenter, and William Bryan, a copperplate-printer, who walked to Avignon in 1789 to attend a society of prophets from all over the world. 'Wright's feet were sorely blistered; but there was no stopping, for his mind was bound in the spirit to travel on.' Fellow-Quietists before their departure, according to Southey, they later rallied to the self-proclaimed nephew of God, Richard Brothers.[23]

Within mainstream literary culture, however, the hardships and privations of traditional pilgrims find but a partial, distant reflection in the self-demeaning pedestrian journeys carried out by the early Romantics. Wordsworth writes intriguingly in 1796 of a plan to travel westwards 'in an humble evangelical way; to wit *à pied*'.[24] Like medieval pilgrims, Romantic pedestrians were participating in a form of symbolic behaviour whereby they disowned the combined authority of birth, family and property, and secular vocation for the duration of their journey.

In Leed's survey of the historical transformations of travel, the 'paradigmatic journeys' of those who have founded religions or civilisations are retraced firstly by insiders and believers (pilgrims) and later by alienated consumers of cultural experiences (tourists). In more general terms, travel in the ancient world, he argues, had a strong connotation of suffering or penance, and pilgrimage did little more than institutionalise this relationship, being 'a formalization of the notion that travel purifies, cleanses, removes the wanderer from the site of transgressions'.[25] While there is little of this conception remaining, in any strict sense, in Romantic tourism, pedestrian travellers in particular were, I have suggested, intent upon clearing an autonomous space for themselves, in which the self could be reduced, physically and intellectually, nearer to its essentials. Certainly other cultural theorists have found it useful to explore the link between pilgrimage and tourism. John Urry puts it thus:

> Like the pilgrim the tourist moves from a familiar place to a far place and then returns to the familiar place. At the far place both the pilgrim and the tourist engage in 'worship' of shrines which are sacred, albeit in different ways, and as a result gain some kind of uplifting experience.[26]

The 'shrines' sought by Romantic tourists were most typically certain varieties of landscape, the visual properties of which they were schooled to appreciate in terms of the sublime and beautiful, and which, in a manner consonant with Urry's notion of the 'tourist gaze', could then be objectified, framed and reproduced via the Claude glass[27] and the ubiquitous sketchbook. Urry argues that the contrast between the ordinary/everyday and the extraordinary/exotic is the fundamental constituent of tourism, and that the desired realm of difference with non-touristic experience is frequently encountered, wittingly or unwittingly, via signs rather than through direct approach to some authentic otherness. In this light, the 'sacredness' of the natural shrines visited by Romantic tourists can be seen as a good deal more fragile than that of the shrines sought by devout pilgrims, since the emotive power and numinous quality of natural landscapes derived in large part from their genuine unfamiliarity to the experience of many tourists, and their accustomising to such scenery arguably led many travellers to become self-conscious, post-touristic consumers of the visual signs of beauty and sublimity. I shall have more to say on this subject shortly.

The sociologist Erik Cohen makes more systematic use of the analogy with pilgrimage in constructing his phenomenological typology of tourist experiences. His classification is organised around different modes of relation between individuals, the spiritual 'centres' of their societies (which may or may not orient individuals' lives in practice), and tourism itself. For Cohen, pilgrimage is defined by a movement from the 'profane periphery' to the 'sacred centre' (within a 'world' that is thought of as larger than the life-space of one's society);[28] tourism, by contrast, involves a movement from the cultural centre (of a secular 'world') to the periphery, or to the centres of *other* cultures and societies. The five modes of touristic experience that Cohen distinguishes, while built on the historical continuity of pilgrimage and tourism, all reinforce these fundamentally different 'social conceptions of space'. Of the five, perhaps the two most relevant to the Romantic travellers I am concerned with are 'recreational' and 'experiential' tourism. Recreational tourism is the normative variety for 'modern man', being a secularised and desublimated version of 'the religious voyage to the sacred, life-endowing centre, which rejuvenates and "re-creates"'; through the pleasures and distractions it provides, it 'restitutes the individual to his society and its values, which, despite the pressures they generate, constitute the centre of his world'.[29] Experiential tourism is the form appropriate to those who are more irreparably alienated from their own cultures, and who seek meaning and authenticity in other places and other ways of life. Insofar as it is a quest for authenticity, it is akin to religious pilgrimage, but differs from it in that the tourist 'remains a stranger even when living among the people whose "authentic" life he observes, and learns to appreciate, aesthetically'.[30] The tone of Hucks's allusion to 'pilgrims performing a journey to the tomb of some wonder-working saint' demonstrates his acceptance of this cognitive distinction, which is manifested more thoroughly in his treatment of the Welsh lower orders as an 'unpolished people' whose 'boldness and originality' and 'love of liberty and independence' he admires, with an alienated eye, in contradistinction to the over-reflective, anxiety-ridden 'commercial people' whose ambivalent representative he is.[31] His detached, 'aesthetic' appreciation of the 'ingenuousness of nature' he finds in the Welsh is typical of the educated sensibilities of Romantic travellers who are prone to seduction by the spectacle (wherever they may find it) of a 'whole' way of life which they can only experience in a vicarious, mediated form.

Equally, Hucks's description of the two undergraduates as 'men travelling for their pleasure and amusement' indicates that their tour partakes substantially of the character of 'recreational' tourism. The blend of recreational and experiential motives that we see in Hucks defines the general character of Romantic walkers as secularised pilgrims: their mode of travel is both a key factor in the pursuit of physical and mental refreshment away from their workaday environments, and a means of enabling their more intimate contact with the real or fantasised realms of authenticity which Romantic writers looked to to reground their over-complex, materialistic societies. Wordsworth's encounter with the pastoral Swiss on his pedestrian tour of 1790, as narrated in *Descriptive Sketches* (which I shall discuss in Chapter 4), is another example of a pilgrim-walker discovering primitivist values on the 'periphery' that he would like to bring back and use to transform his own cultural 'centre'.[32] Curiously enough, however, the Grand Tourists of the eighteenth century, to whose actual motives the published literature is admittedly a fallible guide, share with religious pilgrims an official ideology of personal improvement through travel; it is a paradox that, with the growth of pedestrianism in the Romantic period, tourists come nearer to the physical conditions of pilgrimage when their motives are beginning to diverge in historically significant ways.

THE PHILOSOPHICAL WALKER

Every step now required the greatest care, for even the mere laying hold of a loose stone might have proved fatal. I had once taken hold of a piece of the rock, and was about to trust my whole weight upon it, when it loosened from it's bed, and I should have been sent headlong to the bottom had I not instinctively snatched hold of a tuft of grass, which grew close by it, and was so firm as to save me. When we had ascended a little more than half-way, I was much afraid we should have been doomed to return, on account of the masses of rock over which we had to climb, beginning to increase in size; we knew, however, that a descent would have been attended with infinite danger, and being urged on partly by eagerness in our pursuit, but more from a desire to be at the top, we determined to brave every difficulty. This we did, for in about an hour and a quarter from the time of our beginning the ascent, we found ourselves on the top of this dreadful precipice, and in possession of some very uncommon plants which we had picked up during our walk.[33]

I have quoted at some length from William Bingley's narrative of his ascent of Snowdon in order that the element of surprise, even of bathos, at the end of the passage is adequately conveyed. For several pages Bingley has provided a description of the expedition close and circumstantial enough to encourage one to believe that the danger and excitement he reports are real rather than rhetorically inflated, and to confirm the general impression of his Tour as being in the mould of 'recreational' tourism, with the emphasis on strenuous physical activity, though with aesthetic sidelights and some conventional literary heightening in the form of interpolated poetic quotations. The final half-sentence is almost comically arresting because one's first reaction is one of surprise that Bingley has found time to notice, and to collect, those 'uncommon plants' during a walk in which he was allegedly in fear of his life for much of the way. It is a minor generic discordance that nevertheless gives witness to the often naive overdetermination of Romantic travel writing. Bingley is a recreational walker, and a semi-serious mountaineer, and he aims to convey the rigours and pleasures of his tour with a good deal of autobiographical realism. He also wants to demonstrate that his aesthetic sensibilities are as well-tuned as any other contemporary travel writer's. But he is also a naturalist, and cannot resist trying to arouse the reader's interest in the products of his collector's passion. The problem is that the botanical eye belongs properly to a different discursive tradition, which overlaps awkwardly with the dominant representational practices of Romantic travel literature.

Other pedestrian tourists make more effort to present the different types of discourse woven into their works as a happy interalliance. Adam Walker, from his station above Lake Windermere, conjures up a scene involving particular atmospheric effects, including broken cloud and the massing of light and shade resulting from 'the straight rays, in pencils, streaming before a black mountain'; these, he says, are 'Alpine effects, unseen in flat countries, and afford rational wonder to the Painter, the Naturalist, and the Philosopher'.[34] However, the picturesque gaze assumes a very different mode of relating to the world than that of the naturalist or the natural philosopher, and it is the former that wins out in Walker's tour. Joseph Hucks, whom I have already described as aiming both 'to explore the hidden beauties of nature' and to study 'man in society', veers in the opposite direction, moving most surefootedly in the realm of philosophical abstraction and matters of geographical and historical 'fact', and appearing most uncomfortable at those moments when conventional rhapsodies on natural scenery seem to be required. Thelwall, in *The Peripatetic*, makes one of the most self-conscious attempts to co-ordinate different interests and ways of seeing, in arguing that the pedestrian traveller can turn every situation to account:

> For my part, nothing puts me out of my way; or disturbs the tranquil independence with which I so constantly pursue the beautiful phenomena of nature: a pursuit which no accidental changes of the weather can disappoint: for which of these changes does not produce some additional food for science or imagination? If, as now, a sudden cloud envelope the splendid face of heaven, I compare the appearances with the theories which have endeavoured to explain them. If a storm succeeds, I look around for the shelter of some cottage, or little ale-house, by the way side, where the conversation and manners of an order of society, whose habits, sentiments, and opinions are so widely different from what the usual intercourses of life present me, may pass in entertaining review before me; or, if matters come to the worst, as in the present instance, I accommodate

myself under the shade of some tree, or hovel, and contemplate the operations of nature; see the light mists that had been rarified by the warmth of the lower atmosphere, condensed again by the colder region of the air above, and precipitated in lucid drops to the gaping earth, from which they had formerly been attracted.[35]

Here, meteorological theorising is said to find equal opportunity with studies more appropriate to a social scientist or social historian, and both apparently mesh uncontroversially with the more emotional and imaginative response to 'the beautfiful phenomena of nature' that preoccupies him elsewhere. What is particularly noticeable in this passage, however, is that the interest in men and manners that is serviced by gaining access to the living spaces of a lower 'order of society' expresses itself in the same kind of detached, dispassionate observation ('may pass in entertaining review before me') that is applied to cloud formations and the process of precipitation. Here we sense the enormous weight of the Enlightenment tradition of rational, scientific enquiry and commentary, which bears on a great deal of Romantic topographical and travel writing at the same time as it absorbs the new personae of the 'feeling' observer and the self-realising Romantic ego.

Eric Leed again offers invaluable thoughts on the background of this tradition and the characteristic form of its imprint on the literature of travel. According to Leed, by the fifteenth century curiosity had lost its pejorative connotations and become a legitimating motive of travel, preparing the ground for great scientific expeditions infused with the new empirical creed promulgated by Bacon, whose 'experimental and observational science was the edge of a new theology shriving the senses of their inherited guilt'. As a result of the new importance of travel in furnishing fresh materials for observation, extending the boundaries of experience and knowledge, and cultivating the consciousness appropriate to empiricism, the 'wandering philosopher of the ancient world became the humanist traveller of the Renaissance and the scientific traveler of the seventeenth and eighteenth centuries'.[36] For most of the pedestrian travellers of the late eighteenth and early nineteenth centuries I am concerned with, the 'theology of observation' Leed writes of is an inseparable part of their acculturated mentality, and thus constitutes them in part at least as *philosophical* travellers – or, in my more specific context, 'philosophical walkers'.[37]

This mentality expresses itself most centrally in a facility for observing and recording, comparison, generalisation and abstraction, in a hardening of the outlines of self and world and the precipitation of an 'objective' point of view from the subjective perspectives of the rooted cultural insider. Leed argues persuasively that these intellectual qualities are products of the condition of mobility, which, while conferring advantages in breadth of vision and philosophical 'distance', incurs dangers of superficiality and misunderstanding for which the observer has to compensate by developing a more concentrated awareness and more reflective strategies of interpretation. He further suggests that the prolonged experience of motion encourages what he calls (following Gregory Bateson) a 'progressional ordering of reality' – an attachment to sequence and successivity – rather than a 'categorical ordering of information'.[38] This opposition is reminiscent of the well-known Jakobsonian model of the twin 'metonymical' and 'metaphorical' modes of mental functioning, and subject to similar reservations:[39] certainly the idea of philosophical travel is meaningless without the assumption of the dynamic interrelation of both modes. A typical work of late-eighteenth-century topography such *The Modern Universal British Traveller* shows the dual operation of progressional and categorical methods: each county has a section devoted to it, which moves from a brief natural history, to a very extensive 'topographical description' that deals successively with one major town or city after another (covering historical matters and an account of industrial and commerical activity as well as the more obvious descriptions of place), and finally to a biography of the county which discusses some of its eminent sons. It aims both to guide the 'real Traveller', and to allow the reader's imagination to 'travel with Facility, be entertained on the Journey, gain Improvement at every Stage, survey the whole Kingdom without Danger, and conclude the Excursion informed, but unfatigued'.[40] This seeming indifference to whether the reader is an actual traveller or not is a marker of the scientific aspirations of the work, which exceed the demands of a mere 'guidebook' (a genre that in any case had not made its definitive appearance by this time).

Nevertheless, the distinction between progressional and categorical structures helps to bring into view certain traits and tendencies in the travel literature I am examining. A bias towards the progressional ordering of experience clearly determines the form of much eighteenth-century writing in the genre. Charles Batten, who has comprehensively surveyed the field, says that travel writing required a narrative organisation to be regarded as *literature*, and that this boiled down to the alternatives of journal or epistolary form. These sub-genres are still active in the Romantic period, though a liberalisation of practice takes place on many fronts around the turn of the century. Batten's formalist analysis of eighteenth-century travel writing strikes many chords with Leed's more phenomenological probing of the conditions of what I am referring to as philosophical travel: he draws attention to the Horatian injunction to mix instruction with pleasure; the cultivation of a 'middle' style, avoiding both specialised idiom and colloquial freedom; the elaboration of devices aimed at assuring the reader of the authenticity and reliability of what was described; and the careful intermixture of observation and reflection, with the emphasis on the former and with strict conventions governing the inclusion of the latter (such as conformity to accepted moral or political opinions).[41]

Batten's perceptions of the evolution of travel literature towards the end of the century would suggest a gradual replacement of the philosophical traveller by the aesthetic traveller: he notes a relaxation of the Horatian injunction and the promotion of entertainment values; the incorporation of an increasing amount of autobiographical material (building on Sterne's example in *A Sentimental Journey*) and the accompanying magnification of the writer's subjectivity; and the shift away from conventional topography to descriptions of picturesque beauty, which effected a blurring of the formerly clear line between observation and reflection. While there is ample evidence to support this evolutionary schema, it should not be allowed to conceal the persistence of the practice and representation of philosophical travel, though it is probably the case that as the nineteenth century advances the latter becomes divorced from general touristic experience and institutionalised in the various branches of 'professional' scientific exploration and research. I have already shown the lingering hold which the 'theology of observation' had on early Romantic pedestrian tourists. To what extent their mode of travel can be thought of as assisting or inhibiting the philosophical enterprise is a matter of some doubt: on the one hand, the pace and flexibility of a pedestrian's itinerary might be viewed as providing enhanced opportunities for careful observation and methodical record-keeping, as well as for continuous reflection on what one has seen; on the other, if the rationale of philosophical travel is the ability to make comparisons, to build up a more accurate and less partial picture of the world through observing similarities and differences across larger areas of nature and society, then the pedestrian's very physical immersion in a particular terrain might be considered inhospitable to the generalising, abstracting tendencies of the scientific traveller. But that many pedestrian tourists embraced the discipline of observation, and saw this as one legitimating purpose of their travels, is unquestionable.

Although the mentality of philosophical travel inevitably seems a residual element in the profile of the Romantic walker, something that perhaps belongs more to the official ideology of the classical Grand Tour, there is a sense in which it keys in with the more distinctively Romantic themes of freedom and autonomy which I addressed at the end of chapter 1 and at the beginning of the present chapter. Observation can be an exacting, puritanical theology, but it can also be a self-sustaining passion, and such passion can emancipate. Jeffrey Robinson has a thoughtful passage which suggests how this can occur within a pedestrian context:

> On a walk one is continually encountering the new and, by the 'despotism of the eye', the tyranny of bodily pleasure, willingly forgetting the old. Every forgetting is an assertion of freedom from which the mind goes on another journey. Every forgetting is, in addition, a self-forgetting, an assertion of renewed innocence and pleasure. As we forget, and forget ourselves, we become aware of the gradual fact of hoarding of encounters, impressions, and discoveries. We begin to experience our world as a growing plenitude; the circular imagination is also an autumnal one.[42]

So the constant train of fresh appearances encountered on a walk can endow the mind with its own illusion of novelty or rebirth, and a benign 'despotism of the eye' (a phrase borrowed, of course, from Wordsworth's *Prelude*) can purge all consciousness of past selves and a narrowly defining past life. This almost quantitative theory of personal freedom may seem a far cry from the more serious-minded model of philosophical travel, but it may not be false to the experience of many of those early pedestrian travellers who set off in an eagerness of desire to fill their minds with sensations and perceptions of parts of the world that were geographically and socially alien to them.

AESTHETIC WALKING

> Within half a mile of the top the way became extremely steep and rugged. Here another chasm opens on the left, or opposite side of the mountain, perhaps three times as large as that mentioned above, horrid in the extreme, and here the traveller complains of the narrow and dangerous road, in which, if he misses a step, destruction follows. But he is not bound to venture upon the precipice; the road is safe, and he may every where make choice of his step, for a space of half a mile in width, except within a few yards of the summit, and even there it cannot be less than twelve or fourteen feet wide.[43]

William Hutton, in his unusually realistic account of an ascent of Snowdon in 1799, is tactless enough to blow the whistle on what he sees as the manufactured thrills purveyed by many previous writers of tours. In so doing he may be identifying an inevitable development in the history of landscape response in the eighteenth century. Certainly the pursuit of nature in search of aesthetic satisfactions grew massively in popularity in the last quarter of the century, sponsored by a long-running philosophical debate over the meanings of beauty and sublimity and their application to natural forms; by the cultural esteem accorded to the seventeenth-century Italian landscape artists, Claude Lorrain, Gaspar Poussin and Salvator Rosa; and by the vogue for picturesque travel generated by the writings of William Gilpin. The general history of this transvaluation in perceptions of the uncultivated landscape is too familiar to need lengthy rehearsal here, though I shall offer a more speculative argument in respect of the picturesque in the final section of this chapter. It can be assumed that few pedestrian travellers were immune to the attractions of this form of aesthetic travel.

What is also evident is that the emotions aroused by sublime or beautiful phenomena could easily become, in actuality and *a fortiori* in textual representation, mere automatised reflexes, an inauthentic display of what were held to be appropriate responses. Consider, as one example among many, Adam Walker's description of the approach to Borrowdale in the Lake District:

> Stop at LODORE, and above the Little Inn there is a wonderful view of the Lake, and SKIDDOW as a background. This place

seems a door into BORROWDALE, and almost shut up by a huge overhanging rock, that seems to threaten destruction to the wight who dares to invade a place which nothing but eagles had visited till within this last thirty years.

A little above this, among and over wood-clad Rocks, foams the tremendous Cascade of LODORE! dashing from rock to rock with a hideous roar, that may be heard many miles.... It requires no small resolution to persevere in a visit to BORROW-DALE, when the entrance so powerfully assaults both the eyes and ears; but your courage will be rewarded by a scene of the wildest sequestration that perhaps ever excited human curiosity. The road is by HIGH LODORE to GRANGE, a pretty village, with a ruin on a spiral rock, from whence this Stone Vale is seen to great advantage. Imagination would say, that after the world was finished the rubbish was thrown here![44]

Any doubt that one is dealing here with a very conventional rhetoric is dispersed on a wider acquaintance with travel literature of the period. The so-called 'jaws of Borrowdale' are almost a *locus classicus* of exclamatory sublime astonishment in tours of the Lakes. Thomas Gray had been there over twenty years before Walker, and shuddered at the 'turbulent chaos of mountain behind mountain, rolled in confusion';[45] similar passages could be cited from Thomas West, William Hutchinson, William Gilpin, and Robert Southey, among others.[46] Walker, in fact, does not seem even to have penetrated as far into Borrowdale as West before him; West criticises Gray's exaggerated account of the terrors of Borrowdale, but precisely repeats his impressions of the terrain beyond Seathwaite as party to the 'secrets of the mysterious reign of chaos, and old night'.[47] Exemplifying the remarkable doublethink that characterises this genre of discourse, all these writers recognise in their accounts that the mountains where, according to Gray, 'all farther access is... barred to prying mortals', are in reality regularly traversed by the local dalesmen; the power of aesthetic stimuli, or the pressures of rhetorical orthodoxy, are such as to *virtually* obliterate the features of the living, workaday environment where aesthesis takes place.

Walker's narrative comprehends a number of common elements in the routinised literature of the sublime: epithets favoured for their vague suggestivity ('tremendous', 'hideous', 'huge'), cumulative hyperbolic phrasing ('may be heard many miles', 'so powerfully assaults', 'the wildest sequestration'); particular topoi such as that of the primeval desolation of nature, the 'ruins of creation'; the petrefaction of the observer, stunned into silence; and the idea of a setting that defies or transcends human concerns ('One wonders how the inhabitants could scratch a little bread from among the stones'). One might compare these with James Buzard's inventory of the components of what he calls the 'authenticity effect' sought by tourists in the modern era (the term designates with a certain semiotic knowingness the discovery of essential otherness – what might be summed up as 'Frenchness' or 'the real India', for example – wherein lie the keenest satisfactions of secular travel: he names 'stillness' (the solitary and undisturbed experience of the unique power of a particular place), 'non-utility' (the lack of any felt connection of the place with the kinds of social organisation more familiar to the tourist), and 'saturation' (the sense that a setting is overloaded with significance for the correctly sensitive traveller).[48] Adam Walker experiences a silent awe at what seems to him the 'World's End', and the setting seems saturated both with emotional value and with aesthetic significance, as the enthusiastic allusions to Claude and Salvator in the immediately preceding paragraph make clear; he therefore seems to have assembled the components of a decent authenticity effect.

But to what extent does this 'effect' represent a deception practised upon the reader, or, more to the point, the author himself? As I have stated, Walker's narrative is a mosaic of tired rhetorical formulae, but is it on that account to be viewed as written in bad faith? Given that there must have come a time when the new appreciation of landscape became (for most) a matter of conditioned reflexes, it is nevertheless impossible to draw a line that would separate authentic encounters with natural beauty and sublimity from those artificially induced aesthetic events that can at best only perform the reflexive function of demonstrating possession of a correctly educated sensibility. Buzard's thesis lays heavy stress on the process whereby tourists enter a symbolic circuit, collecting authentic experiences which they then use to 'buy' cultural credibility from a real or imagined audience. Although he notes that there were nineteenth-century critics who objected that the conventional strategies of travel-writers operated to 'bar their ... access to the *real* conditions and cultures through which they were passing' and that these writers 'appeared to let the real slip through their hands as soon as they took up their pens',[49] his own position seems to be that tourists can never transcend their mediated relation to the objects of travel. While this is undoubtedly the safest course for a cultural critic to follow, there has to be a theoretical distinction – even if this distinction is finally impossible to demonstrate textually – between the discovery of new objects of aesthetic experience (such as mountains and waterfalls), around which conceptual frameworks are then constructed, and the later reification of key terms and concepts from these theories to the point where they prematurely compartmentalise, or stand in for, experience, rather than conducting a posterior interpretation of it. It is a distinction between 'real' and 'hyperreal' travel.

Perhaps the issue might also be approached by reference to alternative conceptions of rhetoric, though this will have the effect of complicating the distinction I have just drawn. The world of eighteenth-century travel writing is indeed a rhetorical one: but whereas the idea of rhetoric as a repertoire of formulae, topoi, figures and so on, separate from an independent reality and *lying within the intentional control of the subject*, would encourage the idea that writers like Adam Walker are indulging in a self-conscious display of what might be termed 'aesthetic correctness', the idea of rhetoric associated with antifoundationalists like Stanley Fish, that it is inseparable from belief and that 'the highest truth for any man is what he believes it to be',[50] would instead support the hypothesis that travellers *actually experienced things in these ways*. Take Fish's definition of 'rhetorical man':

> As rhetorical man manipulates reality, establishing through his words the imperatives and urgencies to which he and his fellows must respond, he manipulates or fabricates himself.... By exploring the available means of persuasion in a particular situation, he tries them on, and as they begin to suit him, he becomes them.[51]

It would require only a little rewriting to make rhetorical man a suggestive stand-in for the aesthetic traveller. Exploring the natural world in terms of the rhetoric of beauty and sublimity, the traveller finds that such rhetoric 'suits' him, and thus through manipulating the landscape he re-fabricates himself, *becomes* the aesthetic subject.

I would not wish to claim that the Romantic pedestrian stands in any unique relation to the general phenomenon of aesthetic travel, though clearly the bundle of motivations comprising the latter is a very significant ingredient in the profile of most educated walkers of the period. It is possible to speculate that the peculiar intimacy of pedestrian travellers with the environments they pass through, and the way in which walking fosters a sense of proportion between human beings and the rest of nature unflattering to the anthropocentric fantasies of the former, renders them more susceptible to aesthetic stimuli from landscape, especially to the 'inhuman' might of the sublime, than travellers using other forms of transport that carry an associated feeling of mastery over the surrounding space. But the lack of any substantial textual evidence, at least in Romantic writing, makes one unwilling to press this hypothesis much further: against it, one could point out that the typical sublime moment in travel writing of the period is invariably written from a stationary perspective, whatever the locomotive means otherwise adopted; and recall the evidence produced in Chapter 1 that vehicular transport, particularly in the early part of the period, was little faster than walking anyway. Nevertheless, leaving these thoughts aside, there is one specific branch of Romantic landscape aesthetics, and of aesthetic travel – namely, the picturesque – where it does seem fruitful to explore affinities with pedestrian practice, and it is to this topic that I now turn.

PEDESTRIANISM AND THE PICTURESQUE

Pedestrianism and the picturesque come into vogue together. Pedestrian touring, as I demonstrated in Chapter 1, grew rapidly in popularity in the 1790s, and attained a wide geographical reach in Britain at the same time as Continental travel of whatever kind became less practicable. The cult of the picturesque and of picturesque travel advanced equally swiftly with the appearance in print from 1782 onwards of Gilpin's tours, and its theoretical high-water mark is often taken to be 1794, which saw the publication of Uvedale Price's *Essay on the Picturesque* and Richard Payne Knight's *The Landscape: A Didactic Poem*. Pedestrianism and picturesque travel also involve the same broad sector of society, and, to varying extents, participate in the same democratising trend in cultural life. As far as pedestrianism is concerned, I have already shown how the first generation of recreational walkers was concentrated in the middling orders of society, and this side of the argument needs to be pressed no further. However, some brief comments are required in relation to the picturesque, since this has more commonly been interpreted as a socially conservative aesthetic or as the ideological shadow of agrarian capitalism. Ann Bermingham, for example, makes much of the contradiction between the kind of landscape valued by picturesque theorists and the landscape being created contemporaneously by the agricultural revolution, and argues that the picturesque 'harkened back nostalgically to an old order of rural paternalism', whilst recognising that its sentimental fixation on 'dilapidation and ruin' could also be construed as implicitly justifying modernisation. Alan Liu, who has a lengthy digression on the picturesque in his voluminous historicist reading of Wordsworth, contends that it was incapable of distinguishing scenery from property, and that the picturesque gaze functioned as an imaginative appropriation of land paralleling the actual processes of land development and rationalisation currently at work: enclosing hedges could be incorporated into this gaze because the picturesque 'was itself *visual* enclosure'.[52]

Although this is not the place to enter into what are intricate debates about a very complex movement, it needs to be said that the force and purity of such ideological analysis is proportional to its selective and simplificatory approach to the literature, and that both critics cited are at some difficulty in containing the many cross-currents and ambiguities within the picturesque.[53] To be more particular, the distinction between an actual, physical transformation of the land and a merely imaginary rearrangement of its features would seem to be one worth retaining, rather than blurring, and it is certainly crucial to any attempt to articulate the picturesque with the levelling ethos of pedestrian touring. I am sympathetic to the powerful case made by Kim Ian Michasiw against the kind of seamless relation between discourse and action assumed by historicist critics like Bermingham and Liu. Michasiw opposes the view of Price and Payne Knight as the true theorists of the picturesque, systematising the muddled speculations of Gilpin, arguing instead that their aesthetic creeds have strikingly different orientations that flow from their positions in different strata within the gentry. Whereas Price and Payne Knight were both significant landowners in Herefordshire, actively involved in the management and improvement of their estates, Gilpin was a school headmaster turned country vicar who used the profits from his books for philanthropic purposes within his parish. Price's and Payne Knight's implied readers are fellow-improvers, a small élite with sufficient taste and leisure to bring an aesthetic sense to its relationship with land; Gilpin, by contrast, according to Michasiw, wrote for 'powerless tourists' who were obliged to leave the landscape as they found it, but who could all learn a set of rules that would enable them to recognise picturesque beauty where they found it. Given the inevitable social exclusivity of this order of educational attainment, there was nevertheless genuine equality for those who adopted Gilpin's 'rational amusement',[54] who were likely (Michasiw says) to be men at some remove from the real transformations in the rural economy:

> Beautiful nature, like the beautiful soul, marks a space untouched by a social order for which the lesser gentry was unfit, and to which it responded with a resentful sense of moral and aesthetic superiority.[55]

There is an arresting similarity between this construction of Gilpin's picturesque tourist and the profile I have given of the pedestrian traveller of the 1790s, who was also likely to be a disaffected member of the lower gentry or the adjacent professional classes, or destined for such a place in society. It is worth recalling that when William Combe satirised the vogue for picturesque travel in *The Tour of Doctor Syntax*, he chose for his eponymous hero a poor country curate on thirty pounds per annum – the very social group that is so well represented among the early pedestrians. While it would be misleading to suggest, then, that all picturesque tourists were pedestrians, there seems to be some commonality of spirit between the two, and it should not be surprising to find that pedestrian travellers, among the other motives and desires that made up the composite self I have sketched in this chapter, had a strong mental and emotional investment in the picturesque.

To take this discussion further requires some rhetorical analysis of picturesque literature, and here the common ground between Gilpin, Price and Payne Knight is more important than the differences I have highlighted so far. Both pedestrianism and the picturesque present what might best be described as rhetorics of irregularity. My treatment of the issue of the walker's freedom of movement has provided ample illustration of such rhetoric in the literature of pedestrian touring: walking is promoted as a kind of serial irregularity of movement, as the potential for free improvisation upon rectilinear route-planning. The most cursory acquaintance with picturesque theory will reveal the considerable resonance of irregularity as a concept there too. Perhaps the neatest illustration of this feature, and its convergence with pedestrian experience, is Christopher Hussey's comment on William Shenstone's landscape garden at Leasowes: 'The principle of his garden lay-out was that when a building or other object had once been viewed from its *proper point*, the foot should never travel to it by the same path which the eye had travelled over before'.[56] This strategic separation of foot and eye is made possible because the picturesque landscape expressly encourages the irregular movement across country in which the walker delights.

The pedestrian tourist resists moving in straight lines, and it is precisely a scenery 'free from the formality of lines' – the 'lofty banks' and 'mazy course' of the River Wye, for example – that Gilpin finds especially conducive to picturesque composition.[57] One might also compare the pedestrians' celebration of freedom of movement with Gilpin's more theoretical defence of his choice of picturesque objects on the grounds that they invite the 'free, bold' execution that is a powerful source of pleasure in art.[58] Other students of the picturesque confirm the direct, sensuous pleasure taken in the wandering lines of what Gilpin himself termed 'nature's walks'. Thomas West, taking the view from Latrigg into Borrowdale in his Lakes guidebook, puts it well:

> What charms the eye in wandering over the vale, is, that not one streight line offends. The roads all serpentize round the mountains, and the hedges wave with the inclosures. Every thing is thrown into some path of beauty, or agreeable line of nature.[59]

There thus seems to be a mutual reflection of the picturesque irregularity of lines in nature and the irregularity of the pedestrian's way: the eye discovers picturesque beauty by 'wandering' over the landscape, while the human wanderer who serpentizes round the mountains is essentially at one with his quest, has minimised his intervention in the landscape by impersonating the agreeable lines of nature. Payne Knight's poetic disquisition on the principles of picturesque 'improvement', *The Landscape*, can provide a final example of the interlamination of pedestrian motion with the configuration and texture of particular natural settings:

> Let me, retir'd from bus'ness, toil, and strife,
> Close amidst books and solitude my life;
> Beneath yon high-brow'd rocks in thickets rove,
> Or, meditating, wander through the grove;
> Or, from the cavern, view the noontide beam
> Dance on the rippling of the lucid stream,
> While the wild woodbine dangles o'er my head,
> And various flowers around their fragrance spread;
> Or where, 'midst scattered trees, the op'ning glade
> Admits the well-mix'd tints of light and shade;
> And as the day's bright colours fade away,
> Just shews my devious solitary way.... [60]

Here, whether Knight's images are an ideal inventory of picturesque qualities or derive from the topography of his estate at Downton, the cumulative detailing of the abundance, depth, and uneven texture of the natural surroundings renders the description of the poet's 'devious solitary way' almost redundant.

For Gilpin, roughness – by which he understands properties of both line and texture – is the essential quality of the picturesque, and the main point of difference between the picturesque and the beautiful; rough objects, he declares, allow for most freedom of execution on the part of the artist, and, in terms of form, of colouring, and of the disposition of light and shade, help realise the aesthetic aim of 'uniting in one whole a variety of parts'.[61] Refining Gilpin's theory, Uvedale Price adds the quality of 'sudden variation', producing an anatomy of the picturesque founded on the recognition of variety and intricacy (defined as 'that disposition of objects which, by a partial and uncertain concealment, excites and nourishes curiosity')[62] as the twin sources of pleasure in natural objects.

Both Gilpin and Price, of course, write in the shadow of Burke. The latter's *Philosophical Enquiry into the Origin of our Ideas of the Sublime and Beautiful* had discussed variation as among the properties of beauty, on condition that the variation was gradual and insensible. He brings this within his physiological account of the sublime and beautiful, which distinguishes between those objects or experiences that effect a violent tension or contraction of the nerves and those which cause a relaxation of the body, by noting that sharp angles bring about a 'twitching or convulsion of the optic nerve' that is inimical to beauty.[63] Eroticising this distinction in his inimitable manner, Burke writes of the gentle undulations of a woman's body:

> Observe that part of a beautiful woman where she is perhaps the most beautiful, about the neck and breasts; the smoothness; the softness; the easy and insensible swell; the variety of the surface, which is never for the smallest space the same; the deceitful maze, through which the unsteady eye slides giddily, without knowing where to fix, or whither it is carried. Is not this a demonstration of that change of surface continual and yet hardly perceptible at any point which forms one of the great constituents of beauty?[64]

This is consistent with the overall gender-bias of Burke's account, which, though it does not deny the existence of male beauty, regards beauty, and the 'softer virtues' of kindness and compassion, as the more natural property of women, in contrast to the masculine rigours of the sublime and predominantly male virtues such as justice and wisdom. Burke's willingness to find topographical illustrations of the feelings excited by beauty – he compares, for example, being 'drawn in an easy coach, on a smooth turf, with gradual ascents and declivities' to being 'hurried over a rough, rocky, broken road'[65] – further demonstrates the close complicity between a gendered aesthetics of landscape and a phenomenology of male desire.

When reading Gilpin and Price on the picturesque against the background of their mutual precursor, Burke, one notices that they have found in the picturesque a middle term that articulates the insecurity of definition which occasionally beset the *Enquiry*, and nowhere is this blurring of boundaries more evident than in the erotological dimension of their treatment of landscape. Gilpin states bluntly that the division of objects into sublime and beautiful is no more than a convenience, and that the two are invariably mixed in picturesque art, and his description of the picturesque traveller's pleasure in the mere *pursuit* of his object evinces the appropriate complication of Burke's mapping of beautiful landscape onto the female body:

> We suppose the country to have been unexplored.... And shall we suppose it a greater pleasure to the sportsman to pursue a trivial animal, than it is to the man of taste to pursue the beauties of nature? to follow her through all her recesses? to obtain a

sudden glance, as she flits past him in some airy shape? to trace her through the mazes of the cover? to wind after her along the vale? or along the reaches of the river?[66]

The 'airy shape', the 'mazes of the cover' and the winding vale are perhaps consistent with Burke's inventory of beautiful qualities, but the 'sudden glance' and the 'flitting' passage of the shape are less concordant with the mood of languor and relaxation appropriate to beauty. The confusion is effected by the transformation of Burke's feminised landscape, a woman offered up for languid contemplation, into a neoclassical nymph who teases the male observer's roving eye. Price provides a further interesting variation on this theme. He states that the distinction between the beautiful and the picturesque is most obvious to touch: following Burke, beauty inspires feelings of tenderness and affection and a desire to touch and caress; beautiful landscape invites a correlative sort of caressing by the eye, whereas picturesque landscape, Price asserts, lacks this erotic appeal to the observer. However, he finds it difficult to maintain this position as he picks up Burke's theory of the contrasting physiology of the beautiful and sublime; the leading effect of the picturesque is, in his account, a prurient curiosity:

> it neither relaxes nor violently stretches the fibres, but by its active agency keeps them to their full tone, and thus, when mixed with either of the other characters, corrects the languor of beauty, or the horror of sublimity. But as the nature of every corrective must be to take off from the peculiar effect of what it is to correct, so does the picturesque when united to either of the others. It is the coquetry of nature; it makes beauty more amusing, more varied, more playful, but also,
> 'Less winning soft, less amiably mild.'
> Again, by its variety, its intricacy, its partial concealments, it excites that active curiosity which gives play to the mind, loosening those iron bonds with which astonishment chains up its faculties.[67]

There seems an undeniably erotic dimension to the mental play celebrated here: picturesque nature as coquette offers the lascivious aesthete a pleasurable freedom from the tedium of fixed categories or roles; she provides a 'rougher' encounter than the sensuously supine figure of beauty, without ever threatening the domination of the spectator, whilst also, by virtue of those topographically teasing 'partial concealments', warding off the fear of sublime impotence by prolonging the enjoyments of ocular 'foreplay'. The clear distinction between the beautiful and the sublime dissolves into a generalised erotic attentiveness.

To gather these threads together: I have suggested that in the literature of walking the openness and indeterminacy of the pedestrian's way may be mirrored in the variety and irregularity of line of picturesque landscape. I have also argued that it is in the nature of the picturesque, an intermediate and supplementary category, to point up the non-self-integrity of the opposed values that it bridges: for Price, picturesque roughness – as illustrated, for example, by the thorns of the rose – is the 'fringe of beauty' that demonstrates the interinvolvement of its intrinsic qualities with what is supposedly extrinsic, and which prevents its becoming what it would otherwise be: bald, monotonous and insipid. This kind of self-undoing logic of inside and outside is, of course, a favourite quarry of deconstructive analysis, though we do not need to invoke any theoretical writings of Derrida's to see how picturesqueness questions the assurance of Burkean aesthetics in its interpellation of a male connoisseur who moves between a leisurely appreciation of 'female' beauty and morally bracing Oedipal encounters with the terrors of the sublime. If, therefore, picturesque theory effects a confusion of gender in the aesthetic idea of landscape, what consequences might this have for the pedestrian traveller who looks to that landscape for ratification of his identity within a reciprocal relationship with nature? The overarching sexual symbolism of picturesque travel may well be that of a desiring male subject chasing the 'woodland nymphs' to their 'secret haunts', and 'penetrating' the 'undiscover'd shade' of a dense and variegated landscape;[68] but within this general frame picturesque nature, the object of the 'chase', is a more ambiguous mixture of masculine ruggedness and unrepressed elemental forces, on the one hand, and feminine depths, pleasing variety and partial concealments, on the other. If the pedestrian picturesque traveller is, as I have suggested, at one with his quest – if his rhetorical construction, that is, is similar to that of the picturesque itself – does this quest carry the potential to 'feminise' him? There may indeed be problems of sexual identity for the walker, who refuses the masculine purposiveness of direct progress to his destination, and who relishes instead the teasing meanders and partial concealments of his route. Wordsworth is one obvious example of a traveller who, despite his self-representation as one who is 'free, enfranchis'd and at large', is nevertheless burdened with feelings of unworthiness and unmanliness in such a free-floating existence – leading him, for instance, in the opening of *The Prelude*, to commit himself almost immediately to 'active days' and 'prowess in an honorable field'. Picturesque travel, after all, is not heroic travel, which tests and validates the immutable being of the traveller; nor is it pilgrimage, which purifies the self by stripping it to its essentials; nor is it philosophical travel, which fashions an enlightened self through an agenda of rational exploration and accumulation of knowledge: it is, instead, a form of travel which threatens to disorient the personality of the traveller, since, however much Gilpin promotes it as a rational amusement, it is at least as much an erotic adventure, and therefore shares the destabilising power of those mutual identifications of subject and object that characterise all desire. One reason why the narratives of many pedestrian tourists seem such a montage of discourses is that the personae of the radical walker and the philosophical traveller and other more conventionally masculine selves are constantly being used to check and correct the sensuous excesses of the wandering lover of picturesque beauty.

3
Pedestrianism and Peripatetic Form

So far, in surveying the rise of pedestrianism and analysing the profile of the pedestrian traveller, I have referred chiefly to the prose tours, of varying degrees of factuality, that were the most obvious literary product of the transvaluation of walking in British cultural life. In this chapter I want to shift the focus to poetry, with a rearward look at the eighteenth century, within the context of a broader project of exploring the characteristics of 'peripatetic form' as this emerges in more sophisticated literary works in the Romantic period. However, I want to begin where I left off at the end of Chapter 2, with the picturesque, because there remain some unvisited areas of overlap between that way of looking at the landscape, and the mobile perspectives of pedestrian travel, that provide a useful entry to the aesthetic issues I intend to explore.

The central idea is that of *play*, which is perhaps not an idea that connects easily with a view of the picturesque as a dispassionate assessment of the visual properties of natural scenes, or a tendency to judge those scenes against certain inflexible ideals derived from Italian landscape-artists. Christopher Hussey, whose book on the picturesque is still an invaluable interdisciplinary study, nonetheless cannot resist adopting a satirical attitude to Gilpin's guiding belief that nature, while great in design, is invariably deficient in composition:

> He was thus involved in a perpetual compromise, adapting nature, which he understood only vaguely, to art, which he understood (in his generation) well. He tried to mould what he knew to be above reason, to a rational system. Thus we get perpetually the comical vision of the kindly parson, first abasing himself before nature as the source of all beauty and emotion; then getting up and giving her a lesson in deportment. He saw romantic scenery, and analysed and bottled it into the picturesque.[1]

It is easy, but perhaps unwise, to fall in with Hussey's mockery here, since it seems to misread the characteristic tone of Gilpin's work, which – in keeping with his position that picturesque travel is no more than a 'rational amusement' – takes itself surprisingly unseriously and is quite capable of calling its own bluff.

Probably the most frequently-quoted example of Gilpin's tendency to compare actual scenes unfavourably to picturesque ideals is his suggestion that a mallet 'might be of service' in correcting some well-preserved gable-ends at Tintern Abbey, which 'hurt the eye with their regularity'.[2] For some, this is examplary of his obsessional attachment to abstract criteria of visual propriety and harmony. For others, his rejection of the distant prospect of the Abbey because it shows it 'incompassed...with shabby houses' would be guilty testimony to the picturesque habit of cleansing a scene of every vestige of contemporary social reality in favour of a serene reduction of history to nature (on a closer view of the Abbey, the effects of Time are more visible: 'Nature has now made it her own'). In answer to the second charge, it might be argued that Gilpin's recognition of 'the poverty and wretchedness of the inhabitants' of the Abbey ruins strains the ruminative calm of the picturesque gaze without any help from hostile critics; his forthright statement elsewhere that 'Moral, and picturesque ideas do not always coincide' is given demonstrative force here by his description of the ramshackle home improvised in a 'shattered cloister' by a poor, crippled woman: the scene, intriguingly devoid of any response beyond that of a neutral 'interest', implicitly acknowledges all that is selective, partial and distortive in Gilpin's observational methods.[3] This also helps rebut the first charge, which perversely misreads the signs of the participative, subversively humorous aesthetic project Gilpin recommends to his readers, and is well countered by Kim Ian Michasiw's ironic interpretation of the picturesque tourist's mallet:

> The mallet...is a reminder, like the Brechtian alienation effect, that painting is only painting; that representation reduces, distorts, dissembles; that art and, even more, seeing through art ought to have no privilege and certainly ought not to be the guide in remaking the world.[4]

Michasiw's contention that Gilpin's picturesque rules are a deliberate 'cultivation of artifice' for which no epistemological privilege is sought or claimed, seems persuasive to me, and in line with the social priorities of a man who ranked all his literary activity well below the pedagogical reforms he introduced as a school headmaster and his philanthropic work as a country vicar.

Let us consider another passage in which Gilpin defines and defends the extent of picturesque 'licence':

> Such alterations only your artist should make, as the nature of the country allows, and the beauty of composition requires. Trees he may generally plant, or remove, at pleasure. If a withered stump suit the form of his landscape better than the spreading oak, which he finds in nature, he may make the exchange – or he may make it, if he wish for a spreading oak, where he finds a withered trunk: He has no right, we allow, to add a magnificent castle – an impending rock – or a river, to adorn his foreground. These are *new features*. But he may certainly break an ill-formed hillock; and shovel the earth about him, as he pleases, without offence. He may pull up a piece of awkward paling – he may throw down a cottage – he may even turn the course of a road, or a river, a few yards on this side, or that.... Most of these things may *in fact* be altered to-morrow; tho they disgust to-day.[5]

At this point the unregenerate Marxist critic would wade in and condemn the aesthetic freedom Gilpin celebrates as hand in glove with the physical despoliations of tyrannous landowners who laid waste villages in order to enlarge their pleasure gardens. But visual appropriation is not the same as physical appropriation, and the picturesque tourist who razes a cottage to the ground, or turns the course of a road, in a sketch or in the mind's eye is not on a par with the monopolising landlord or the enclosure commissioner: the satisfaction he finds in aesthetic play alters nothing and displaces nobody – the only displacement such play effects is of the mental representation of the landscape-in-itself, which for Gilpin, as Michasiw rightly notes, is essentially unrepresentable. Rather than a surrogate form of usurpation, it functions as a discreet and unsubstantial compensation for the social disempowerment of the average picturesque tourist, a language game that brazenly displays its independence from the discourses of social and political action.

In discussing the 'sources of amusement' appertaining to picturesque travel, Gilpin lays great stress on certain mental operations that are both playful (in that the pleasure they furnish is noninstrumental) and 'scientific' (in that they proceed through analysing scenes into their component parts, comparing these parts with similar objects and appearances elsewhere, and so on). In addition to the satisfaction derived from a more comprehensive 'knowledge of objects', there are the various pleasures of representation: these include the practice of making sketches, which 'serve to raise in our minds the remembrance of the beauties they humbly represent', and which flatter 'with the idea of a sort of creation of our own'; and the production of 'scenes of fancy', which is the more genuine creative employment of the picturesque sensibility, and which consists in selecting and combining images from nature according to the 'rules of art', and 'in the best taste'.[6] Gilpin adds an interesting paragraph on the pleasures obtained from the purely internal 'visions of fancy':

> Often, when slumber has half-closed the eye, and shut out all the objects of sense, especially after the enjoyment of some splendid scene; the imagination, active, and alert, collects it's scattered ideas, transposes, combines, and shifts them into a thousand forms, producing such exquisite scenes, such sublime arrangements, such glow, and harmony of colouring, such brilliant lights, such depth, and clearness of shadow, as equally foil description, and every attempt of artificial colouring.[7]

Such imaginative play, Gilpin is at pains to make clear, in no way departs from the 'simple standard of nature': the imagination improves upon nature using principles derived from nature itself. But I want to emphasise the level of experimentation and intense re-envisioning ascribed to the picturesque eye, which has often been superficially conceived as casting a fixed, inert gaze, drily re-composing each perceived landscape in accordance with certain invariant laws of aesthetic form. It is true that the 'exquisite scenes' presented to the mind's eye are still recognisably sanctioned by codes of visual appreciation that could be brought to bear with equal legitimacy on a hundred other localities, but the kind of play celebrated is in excess of that required by a purely inflexible system. This liberty-within-confinement, exercised in particular when the picturesque eye relaxes and looks inward, has points of contact with the conditions of pedestrian travel, and the properties one might expect of an aesthetic grounded in the mental and bodily experience of walking, that are worth considering.

My line of argument here is determined partly by disagreement with Anne Wallace's severer judgement on picturesque travel in the course of defining her own idea of a peripatetic aesthetic. For Wallace, Gilpin's methods are entirely consistent with the theory and practice of what she terms 'true travel', which subordinates passage to destination, promotes the 'continuing, stabilizing isolation of place from place', and commits itself to the exact observational science I discussed under the heading of 'philosophical travel' in Chapter 2. Although Gilpin's approach might seem remote from the fidelity of description sought by topographers, explorers and such like, Wallace argues that

> Gilpin simply extends objectification to a complete detachment of parts from contexts so that, by reassembling the parts into more desirable configurations, fidelity to artistic (rather than scientific) principles can be achieved. To an even greater extent than the scientific traveller, in fact, the picturesque traveller avoids dangerous psychological engagement in movement, process, and change, for he need not perceive or represent his destinations as they are, but may immediately reshape them to fit his already settled ideas of beauty.[8]

Strangely enough, a few pages earlier Wallace discusses the maintenance of strict spatial and temporal separation between observations of different places at different times as a major ground of distinction between 'the aesthetic of true travel and that of peripatetic', thereby investing the latter, which she elsewhere describes as a 'redrawing the world by discourse', with the properties of recombining and coalescing originally distinct images and ideas.[9] It would seem that Wallace's modelling of peripatetic theory upon a particular kind of recollective, synthesising labour that she finds identified with 'excursive walking' in a sector of Wordsworth's mature poetry, has prevented her from seeing those aspects of the picturesque sensibility that are congruent with the perceptual and intellectual processes characterising pedestrian travel. In fact, the picturesque tourist as constructed in Gilpin's writings seems very sensitive to 'movement, process, and change', and the mental play – the transposing, combining and shifting – he writes of proceeds from, is the aesthetic elaboration and intensification of, this experience of movement and change. All travel stimulates a comparative consciousness, and the picturesque tourist is no different to any other traveller in having developed a facility for making comparisons through the mobility s/he enjoys; equally, the imaginative play Gilpin relishes is the product of having experienced a constantly changing landscape at a certain pace, a pace which allows for the details of natural scenes to be impressed upon the mind, which permits the lingering gaze and the backward look, as well as providing the freedom to stop and rest at will, and which creates sufficient space for the interaction of reverie with the objective world. Pedestrianism is not the only mode of travel that meets these conditions, but it *does* meet them, and this indicates the falsity of positing a hard line between the picturesque and peripatetic aesthetics. Whilst I would repeat the obvious point that not all picturesque travellers were pedestrians (a slow horseback was Gilpin's own preferred mode of conveyance), it is equally true that many of the pedestrian tourists I have discussed had a trained picturesque eye, and the apparent incongruity between this aesthetic orientation and their means of locomotion largely disappears on closer inspection. Picturesque theory, in fact, is of some help in formulating an inventory of the traits one might expect to find in the aesthetic practices developed by the more talented and individual writers I shall consider in Chapters 4–7, who all give purposeful expression to the ways in which pedestrian motion can condition or mediate thought and perception.

The best way to approach this task is to establish some of the distinguishing features of walking not so much in general terms as one mode of travel opposed to others, but more specifically as a way of experiencing landscape or as a form of consciousness-in-motion. Among these features, which should readily be recognised by anyone who has done any long-distance or endurance walking, are the following:

(i) The pedestrian's experience of landscape is a *participatory* rather than disinterested one: s/he is in constant sensuous contact with the environment and is entirely responsible via voluntary movements of the limbs for what s/he perceives of the (natural) surroundings. The achievement of any 'view', or any final or intermediate destination, is inseparable from the physical exertion which it required. A study of highway aesthetics points out that 'The modern car interposes a filter between the driver and the world he is moving through. Sounds, smells, sensations of touch and weather are all diluted in comparison with what the pedestrian experiences'; enhanced kinaesthesia compensates the driver for this impoverishment of sense.[10]

(ii) The pedestrian's experience of the world is of a slowly but continuously changing field of appearances. Eric Leed writes of the 'perceptual envelope' inhabited by the passenger, in which the only invariants in a 'flowing perceptual array' are the 'aiming point' towards which one is heading, and from which objects emerge and grow in size as one approaches, and the 'vanishing point' into which things dwindle in size behind one. Leed suggests that this envelope 'elongates into a tunnel as the speed of passage increases', and that when such travel is sustained a 'flow state' can be induced in which distinctions between self and world, and between past, present and future are elided. However, a main characteristic of this flow state is the suspension of volition, in that it 'allows the passenger not to think, or to think according to the order in which appearances present themselves', and this seems ill-suited to the experience of pedestrians, whose feet are under their own control and who have constantly to think about where they are placing them to avoid falling flat on their faces.[11] Most pedestrian travel is undertaken at a moderate, steady pace – in a broader, more finely-grained perceptual envelope – that provides complete freedom to stop and restart at will, and allows for the lingering gaze and the backward look; it offers a subtly altering panorama or vista in which the same objects are seen and heard from different heights and angles and distances.

(iii) Despite the volitional character of walking, on a long-distance walk the regular, alternating rhythm of right leg, left leg, can induce a hypnotically self-absorbed state (if the conditions of the ground are not such as to demand constant vigilance), which one typically moves into and out of, and which can take the form of an irregular mental play – the wandering of reverie, or the re-presentation to consciousness of what has been seen and felt – as well as more abstruse forms of introspection and concentrated creative thought. This enhanced mental excitation is possible because walking has a remarkable ability to purge the mind of its habitual, everyday clutter. Richard Long, the artist and sculptor who has made long-distance walking the very medium of his work, has said that 'getting myself into these solitary days of repetitive walking or in empty landscapes is just a certain way of emptying out or simplifying my life', and that 'having the rhythmic relaxation of walking many hours each day puts me into a state of mind which frees the imagination'.[12]

(iv) The early twentieth-century travel writer Fillippo de Fillipi was inspired by the Karakoram mountains to write: 'Walking is really the only kind of locomotion that puts us on equal terms with the world about us. Our modern mechanical methods of transportation tend to make us lose sight of our relative importance.'[13] It would be going too far to proceed from this unexceptionable statement to an assertion that long-distance walking inevitably distils some kind of 'ecological' consciousness, but a readjusted sense of proportion between humanity and the wider natural environment is a commonly reported outcome in the literature of serious walking. The pedestrian *at best* feels 'equal', and often feels *un*equal, to his/her surroundings; for all the freedom of movement and independence it makes available, walking never permits that illusion of the personal mastery of space enjoyed by the motorist, the motorcyclist, the pilot, or even the skier.

(v) Because of the moderate pace of his/her itinerary, and the intimate sensory contact with the environment that walking presupposes, the pedestrian is more alert to the multiplicity of appearances and the particularity of actual landscapes. Walking, in other words, is capable of fostering resistance to any idealising aesthetic tendencies the traveller may start out with, and of countering the generalising and abstracting mentality inherent to all travel (as described in Chapter 2).

(vi) The pedestrian tourist, or any other practitioner of long-distance or outward-bound walking, experiences the world, like seasoned travellers of other kinds, as a sequence or sequences of towns, villages, rivers, roads, paths, hills, vallies, sights, sounds, and so on. Although the mind may strive to organise the resulting impressions and memories in a variety of more conceptual ways, the dominant mental set is towards what has already been described as the 'progressional ordering of reality'. A fascinating analogy to the linear modelling of space that this entails is to be found in the Australian Aboriginals' 'songlines' investigated by Bruce Chatwin: the Aboriginals, he discovers, 'could not imagine territory as a block of land hemmed in by frontiers: but rather as an interlocking network of "lines" or "ways through"'. These lines are identical with the songs by which their Ancestors sung the world into existence in the Dreamtime, and which have to be ritually re-performed to prevent the land dying: each man owns 'his stretch of the Ancestor's footprints', which he must sing 'Always in the correct sequence'.[14] For the Aboriginals, 'home' is an elongated pedestrian itinerary, intersected by other men's homes; for the pedestrian tourist in the modern era, 'home' is the settled, bounded location from which one departs, but the journey that follows shapes a mentality governed just as powerfully by sequence. Moreover, the irreducibility of the line discourages the search for unity and completeness: an acceptance, rather than transcendence, of change and difference is the more likely reward or goal of the committed walker.

It would be foolish to expect that all these qualities of 'the mind of the walker' will be comprehensively encoded in every piece of walking literature from the Romantic period onwards. Nevertheless, it is arguable that these traits begin to find rhetorical-poetic expression, in more partial and sometimes inexplicit ways, as one nears the end of the eighteenth century. One might elaborate on this by drawing on George Santayana's distinction between the sensory, formal and expressive levels of aesthetic experience.[15] On the first level, one would expect that the pedestrian's unique exposure to sensory stimuli from the country s/he passes through would give rise to more particularised and 'realistic' observations of nature, in certain kinds of writing, as the middle-class cultural reclamation of walking takes place: one might look, that is, for more strenuous attempts to explore and record the individuality of different locations (as against their conformity to prefabricated landscape ideals), as well as the consciousness of the perceiving subject who is, in the fullest sense, as a walker, a 'material witness' of their ever-changing appearances. On the formal level, one is concerned with the persisting application of formal perspectives, such as the pictorial, to natural landscapes – an aesthetic tendency that seems at odds with the less heavily mediated realm of sensory aesthetics, but which must take account of the many congruences between the picturesque and pedestrian travel I have elaborated. Beyond this, there are other formal characteristics of walking – irregularity of line, seriality and progression, non-synthesizability, and so forth – that one might expect would have aesthetic repercussions in some of the more sophisticated versions of peripatetic. Santayana's 'expressive' level deals with the symbolic properties of objects, including those which have a long history of cultural reproduction and transmission: it is in this light that one might consider the 'global' metaphorical uses that have been made of walking in the literary tradition, and its assimilation to more general figures, such as happens within the enormous intertext of the life-as-journey allegory. These kinds of symbolism have little to do with the material practice of walking, but one would expect to see them re-enlivened in Romantic literature as that practice emerges from the closet of social-political disdain and containment.

I am making (or rather, accepting) no case for the theoretical purity of these three levels of aesthetic experience. Indeed, I have provided ample justification already in this book for suspecting their integrity: the idea of 'aesthetic correctness' I explored in Chapter 2, for example, was predicated on the difficulty of distinguishing between the emotions produced by raw sensory experience and the manufactured delight and horror accessible to literary sensibilities (wherein a rugged mountain, say, is not a *cause* of sublime astonishment but a conventionalised *sign* of such emotion). Nevertheless, Santayana's categories help construct a rough framework of expectation for assessing the transformations in the representational life of walking that are a decisive factor in some important Romantic texts, and in the remainder of this chapter I would like to build on that framework by surveying some revealing eighteenth-century antecedents.

One useful way of measuring the changes I am talking of is to track the line of evolution embracing three poems that share the basic device of the 'prospect-from-a-summit', a sub-genre within the broader field of topographical poetry: Sir John Denham's *Cooper's Hill* (1642), John Dyer's *Grongar Hill* (1726), and William Crowe's *Lewesdon Hill* (1788). The hill which provides Denham's assumed poetic vantage-point lies about eighteen miles from London, and enjoys a panorama that includes (or did in the seventeenth century) Windsor Castle and St Paul's Cathedral, both of which Denham makes the subject of extended reflections. However, he swiftly makes clear, in his opening apostrophe to Cooper's Hill, that he does not intend to be limited by the physical restrictions of his present position:

> Nor wonder, if (advantag'd in my flight,
> By taking wing from thy auspicious height)
> Through untrac't waies, and ayrie paths I flye,
> More boundlesse in my Fancy than my eye.... (ll. 9–12)[16]

Denham has invisibilised the entire process of climbing Cooper's Hill, and he is no sooner there than he is darting off along the 'ayrie paths' of imagination: flight is the appropriate metaphorical vehicle for the progress of his poem, which deals in a largely idealised landscape insofar as it deals in landscape at all, and in which moral, historical and political reflections vastly outbulk the descriptive and circumstantial elements (which provide light-weight 'prompts' for Denham's meditative excursions). The Thames, for example, catches the poet's eye, 'descending from the hill' (l. 159), and is immediately converted to an emblem of political moderation and balance and a metonymy for English trade and exploration ('to us no thing, no place is strange, / While his fayre bosome is the world's exchange'); while the topography of the Thames valley, which includes an 'ayery Moun-taine' difficult to identify with leafy Surrey, is made representative, in all its generalised perfection, of the 'unity in variety' of God's creation:

> While drynesse moysture, coldnesse heat resists,
> All that we have, and that we are, subsists.
> While the steepe horrid roughnesse of the Wood
> Strives with the gentle calmnesse of the flood.
> Such huge extreames when Nature doth unite,
> Wonder from thence results, from thence delight.   (ll. 207–12)

The most vivid section of the poem is the long description of a stag hunt in lines 241–322, but this is presented as recollection rather than observation: the allegorisation of the fate of the stag, who 'disdaines to dy / By common hands', always seems imminent, and is applied explicitly to the bringing-to-bay of the king at the time of Magna Carta, though the subtextual reference to the vicis-situdes of Charles I is plain enough. In sum, Denham's elevated situation on Cooper's Hill is neither won by any physical effort nor positioned in any linear travel sequence, but rather is a rhetorical site from which a heavily moralised landscape is made to yield up its lessons of cosmic and political order.

Dyer's *Grongar Hill* represents a marked development from *Cooper's Hill* in terms of the actualisation of the landscape and of the poet-observer's participation in it. The summit itself is gained almost as painlessly as in Denham's poem, but although there is no attempt to describe the changing vistas of the climb in any detail, the general sensation of perspectives changing with an increase in height is scrupulously recorded:

> The Mountains round, unhappy Fate,
> Sooner or later, of all height!
> Withdraw their Summits from the Skies,
> And lessen as the others rise:
> Still the Prospect wider spreads,
> Adds a thousand Woods and Meads,
> Still it widens, widens still,
> And sinks the newly-risen Hill.   (ll. 33–40)[17]

However, if these lines suggest deference to the perceptual realities of pedestrian movement, the bulk of the poem's landscape description shows the same idealising tendency as *Cooper's Hill*: it is relentlessly nominal, with most objects allotted only typological attributes ('The gloomy Pine, the Poplar blue, / The yellow Beech, the sable Yew' [ll. 59–60]), inclines towards the pastoral ('While the Shepherd charms his Sheep, / While the Birds unbounded fly, / And with Musick fill the Sky. / Now, ev'n now, my Joy runs high' [ll. 142–5]), and evinces a delight in the variety of forms similar to Denham's, though now in the service of a pictorial aesthetic modelled on the paintings of Claude and Salvator:

> The windy Summit, wild and high,
> Roughly rushing on the Sky!
> The pleasant Seat, the ruin'd Tow'r
> The naked Rock, the shady Bow'r;
> The Town and Village, Dome and Farm,
> Each give each a double Charm,
> As Pearls upon an *Aethiop's* Arm.   (ll. 106–13)

Immediately after the lines above, Dyer initiates what seems a more particularising description, one that might mark a movement towards endeavouring, in John Barrell's words, 'to localise landscape, to represent, not *landscape*, but *a landscape*':[18]

> See on the Mountain's southern side,
> Where the Prospect opens wide,
> Where the Ev'ning gilds the Tide;
> How close and small the Hedges lie!

> What streaks of Meadows cross the eye!
> A step methinks may pass the Stream,
> So little distant Dangers seem.... (ll. 114–20)

However, just when it seems that Dyer may be ready to elaborate on the peculiarities of visual appearances, he instantly turns to moralise the last observation into a warning of the uncertainty of worldly things and the treachery of 'Hope's deluding Glass' (122). One is left with the impression that, if the 'message' does not exactly manufacture the observation, Dyer's respect for the laws of genre is such that when he scrutinises a landscape which he must, in fact, have known very well, he selects only those details that can be brought under a conventional moral-aesthetic agenda.

Grongar Hill also features in another of Dyer's poems, 'The Country Walk', which, with its ostensible pedestrian theme, makes an interesting comparison with the 'prospect-from-a-summit' piece. There is more of a progressional ordering of appearances in this poem, with some details at least eluding the long arm of the poetic moralist: the description of 'A beautiful variety / Of strutting cocks' in front of a 'yellow barn', inside which 'rustics thrash the wealthy floor',[19] may be rather too charming to be convincing, but there follows no Thomsonian panegyric on Britain as the 'exhaustless granary' of the world,[20] and the threshers at least manifest more physical vitality than the shepherd later found 'Leaning on a bank of moss' in a routine posture of pastoral indolence. Similarly, an old man observed puffing upon his spade and digging up cabbage is neither lauded as an emblem of humble rural industry nor openly objectified in his essential picturesqueness, but has his features and movements plainly and neutrally recorded.

This is, however, as far as Dyer proceeds with the realism of his country walk. The chronicling of pedestrian motion largely takes place through regular variations on the same sparse formula: 'And now into the fields I go...A little onward, and I go...I rouse me up, and on I rove.' This has the effect of foregrounding the rhetorical structure of the poem at the expense of diminishing the legibility of the walker's experience. When the poet-observer ascends with customary rapidity to the 'bushy brow' of Grongar Hill, the prospect he obtains is as departicularised as that enjoyed in the companion-text. The most repetitive trait of this generalising vision, which runs through the whole poem but is arrestingly compacted in the summit view, is the pluralising of concrete nouns in brisk serial formations:

> Where am I, Nature? I descry
> Thy magazine before me lie!
> Temples! – and towns! – and towers! – and woods!
> And hills! – and vales! – and fields! – and floods!

The singular poverty of adjectival qualification in the poem confirms its main thrust as being the representation of an ideal abundance and harmony, a non-specific variety in which the mere concatenation of objects is more important than their precise local character. The country walk is the structural device by which Dyer effects this display of rural harmony, a neighbourly co-existence of humble poets, hoary old men and jocund shepherds, with the only ineffectual trace of a hierarchical social order being a safely dilapidated 'remnant of a seat' on the hill opposite. In terms of the inscription in poetry of pedestrian passage through a natural landscape, Dyer's poem therefore marks a minor yet significant deflection of the topographical tradition, with established codes and conventions constraining the tendency to a more naturalistic treatment of the mind of the walker.

Reading William Crowe's *Lewesdon Hill* against the backdrop of this tradition, one is hardly surprised to find the poet rapidly transported to the top of the hill, there to survey a typically 'variegated scene, of hills / And woods and fruitful vales and villages / Half hid in tufted orchards'.[21] Some of Crowe's miscellaneous reflections would not have been discordant on Cooper's Hill in the seventeenth century: the disappearance of Glastonbury Tor in mist, for example, incites a meditation on the illusoriness of earthly things and the 'havoc of wide-wasting Time' (p. 36). However, Crowe's poem, loose and digressive in the best georgic-topographical tradition, conveys a more naturalistic sense of actual time and place than his predecessors. The May morning he describes is no bucolic ideal, but a time verified with homely detail of green foliage and flowering hawthorn. The many Dorsetshire place-names he weaves into the poem (as belonging to settlements or landscape-features within his view: Burton Cliff, Nethercombe, Sherberne, and so on) not only reinforce the impression of concrete locality, but allow the landscape to be invested with a measure of historical depth, since the reflections they inspire are not general moralisations but are matters of local knowledge and popular legend.

The most significant development in *Lewesdon Hill* is the proto-Romantic emphasis on inner space, and on the walk and the landscape as medicine for a mind oppressed with worldly cares. On first attaining the summit of Lewesdon hill Crowe feels 'the mind / Expand itself in wider liberty' (p. 6); then, in what seems a revisionist glance at Denham's Thames, the mighty river of trade ('Full of the tributes of his grateful shores' [*Cooper's Hill*, l. 182]), he finds in a 'nameless Rivulet' that flows from Lewesdon to the sea 'Untainted with the commerce of the world', an objective correlative for the 'free unbroken spirit' (pp. 11, 7). He attempts to give this assertion of free subjectivity some topical philosophical substance in a long passage attacking necessitarianism, though his argument amounts to little than a grand tautology: 'that the mind is free, / The Mind herself, best judge of her own state, / Is feelingly convinced' (p. 15). In fact, the closing lines of *Lewesdon Hill* expose the confined territoriality of the unbroken spirit, which declines in self-confidence as Crowe descends the hill:

> Now I descend
> To join the worldly croud; perchance to talk,
> To think, to act as they: then all these thoughts,
> That lift th'expanded heart above this spot
> To heavenly musing, these shall pass away
> (Even as this goodly prospect from my view)
> Hidden by near and earthy-rooted cares.
> So passeth human life – our better mind
> Is as a Sunday's garment, then put on
> When we have nought to do; but at our work
> We wear a worse for thrift.   (pp. 39–40)

It seems that the mind is free, but only on a Sunday, when it is let out for some strictly timetabled recreation: Crowe's simile comparing his 'better mind' to Sunday clothes is an unfortunate one, since it connotes a kind of genteel externality that militates against his professed epistemology. The poem therefore finishes by affirming a fugitive selfhood that defines itself against, but has difficulty extricating itself from, the functional mentality of quotidien life. It is a feebly delineated Romantic predicament, but a recognisable one, and in the context of my discussion it is significant that it takes as its vehicle the liberating effects of a more naturalistically rendered pedestrian excursion.

The topographical line from Denham through Dyer to Crowe shows the gradual transformation of the walker from an organising trope encumbered with political, moral and aesthetic agendas of an adventitious or externally validated nature, to a more material figure-in-the-landscape possessed of an unfolding selfhood whose health and identity are bound up with the excursionary experience itself. This latter figure is, of course, no less rhetorically constructed than the former, but its rhetoricity is more in tune with received ideas of the structure of modern subjectivity, and therefore seems more natural and offers easier passage to the reader's sympathies.

A few more examples from eighteenth-century literature will help to demonstrate the weight of the literary traditions from which Romantic pedestrians, in the course of their real-life emancipation from the class opprobrium directed at walking, had to detach themselves as they represented their activities in writing. Oliver Goldsmith's *The Traveller, or a Prospect of Society* (1764) is a good example of the tradition of philosophical travel I sketched in the previous chapter. The poem reflects his experiences on a Grand Tour undertaken in 1755, and aims to show, in harmony with the official ideology of such tours, 'that there may be equal happiness in states, that are differently governed from our own; that every state has a particular principle of happiness, and that this principle in each may be carried to a mischievous excess'.[22] In the opening lines he compares his brother's contented, settled and useful life with his own itinerant existence, 'Impell'd, with steps unceasing, to pursue / Some fleeting good, that mocks me with the view' (ll. 25–6). However, for the purposes of the poem these unceasing steps are exclusively cerebral, for the Traveller immediately takes a rhetorical seat high in the Alps and begins a philosophic prospect of 'an hundred realms' and their various claims to have monopolised 'Creation's charms' (ll. 34, 37). He exults in his panoramic view of all the good 'that heaven to man supplies', but still hungers after some more favoured spot 'Where my worn soul, each wand'ring hope at rest, / May gather bliss to see my fellows blest' (ll. 56, 61–2).

The objectivity and comprehensiveness of mind which this aerial contemplation presupposes, and which were lauded as the hard-won outcome of extensive experience of other cultures by the advocates of philosophical travel, are, in fact, in place at the start of the poem and control every moment of its progress through the countries of western Europe. Despite the actuality of Goldsmith's travels on the Continent, there is no interest in representing the mental transformations effected by a prolonged period of open-ended travel. Instead, having surveyed the natural advantages of a number of countries and the merits and demerits of their national cultures (including England, whose freedom is celebrated, but whose excessive love of independence is found to weaken the social bond), his travels are found to have been vain and circular, since, given that 'reason, faith and conscience' are inalienable, mental and spiritual contentment can be found under any system of government:

> How small, of all that human hearts endure,
> That part which laws or kings can cause or cure
> Still to ourselves in every place consigned,
> Our own felicity we make or find.... (ll. 429–32)

The structure of Goldsmith's poem is therefore that of a circuit of Western Europe designed to quell the immoderation of his 'wandering hope' (l. 61) and restore the serenity of the 'philosophic mind' (l. 39). Although he presents himself as one whom 'fortune leads to traverse realms alone' (l. 29), the experience of solitary passage is scarcely rendered in the text, which inclines to trope its own progress locomotively as flight: 'To men of other minds my fancy *flies*' (l. 281; my emphasis), is the transition to a section on the industry and avarice of the Dutch, while Goldsmith's 'genius spreads her wing' (l. 317) on turning to assess the 'lords of human kind' (l. 328) in Britain. The process of the poet's wandering steps is invisibilised because the main thrust of the poem is the moralising of the journey considered as an ordering of discursive topoi on the wealth of nations, and this does not require that such travels have actually taken place.

William Bowles also undertook extensive travels on the Continent (and, unlike Goldsmith, undertook them on foot), but finds it difficult to give life to these travels because the *Sonnets* that grew out of his tour are so easily co-opted by the traditional metaphorical appropriations of *any* journey as a commentary on the progress of life or an expression of some received spiritual teleology. Sonnet 2 glosses the 'languid' traveller's habitual backward glance as the inevitable nostalgia of the disillusioned adult for the 'pleasing prospect of the past', as opposed to the beguiling vista of life's onward journey at a time of 'vain hope'. Sonnet 6 finds Bowles's 'wand'ring feet' treading the banks of the Tweed, but characteristically it is not his feet but his 'thoughts' that are 'weary': his pedestrian exertions are sublimated as the torments of a sensibility maladapted to the 'stormy world'. Overall, Bowles's journey is denarrativised into a loose assembly of discrete moments of stationary meditation: the linearity and progressionality of the tour are legible only via selected points of interrruption, while movement through space is assimilated to the poem's sense of time (and thereby obscured) in Bowles's near-ritual acknowledgement of 'the hour that stays not'.

William Cowper's long poem *The Task* (1785) is rightly highlighted by Anne Wallace for marking a movement towards a 'peripatetic aesthetic', however 'entangled' it remains in 'the desire for effortless panorama'.[23] Certainly the tension between picturesque 'stations' and a more progressional experience of landscape is one of the most striking features of the poem, especially in Book I. Cowper moves unhurriedly from his mock-heroic preamble on the sofa to a disquisition on his physical fitness (his retention in middle age of 'The elastic spring of an unwearied foot / That mounts the stile with ease'),[24] to a description of a habitual walk to the top of an unnamed 'Eminence':

> Thence with what pleasure have we just discerned
> The distant plough slow-moving, and beside
> His labouring team that swerved not from the track,
> The sturdy swain diminished to a boy!
> Here Ouse, slow-winding through a level plain
> Of spacious meads with cattle sprinkled o'er,
> Conducts the eye along his sinuous course
> Delighted. There, fast rooted in their bank,
> Stand, never overlooked, our favourite elms,
> That screen the herdsman's solitary hut;
> While far beyond, and overthwart the stream,
> That, as with molten glass, inlays the vale,
> The sloping land recedes into the clouds.... (I. 159–71)

Although the zealous editor of my Aldine edition of Cowper notes that these lines 'contain a literally accurate description of the leading objects which meet the eye on a walk westward by a pathway over fields from Olney to Weston', it is transparent that they do not represent these objects in a manner imitative of their occurrence on a pedestrian excursion: rather, Cowper presents a stationary prospect with scrupulous attention to middle distance (the ploughing team and herdsman's hut) and background (the distant hills and low clouds, towards which the slow-winding Ouse leads the eye in the manner of the diagonal link favoured by Claude and others).

Against this predilection for the picturesque gaze, one might place the slightly later passage fastened upon by Wallace, in which Cowper brings to life another local excursion as a fully bodily encounter with the landscape, walking 'ankle-deep in moss and flowery thyme', and feeling 'at every step / Our foot half sunk in hillocks green and soft' (I. 270–2); or the engaging passage at the start of Book 5 (entitled 'The Winter Morning Walk') in which he focuses upon his own shadow cast by the horizontal early sun:

> The shapeless pair,
> As they designed to mock me, at my side
> Take step for step; and as I near approach
> The cottage, walk along the plastered wall,
> Preposterous sight! the legs without the man. (V. 16–20)

Or there are later parts of the description of Cowper's walk to Weston in Book 1 that give some sense of the 'perceptual envelope' inhabited by the pedestrian traveller. The boundaries of this envelope are slowly but constantly shifting to give the walker's experience its particular sequential character, and since there is no physical interference (carriage, etc.) between walker and landscape the data are not exclusively visual, though inevitably the enjoyments of the eye predominate. The following passage underscores too the politics of perception, as Cowper notes that 'folded gates would bar my progress now' (I. 330), and thereby registers the permissive nature of his walk across an owned and managed landscape. The 'guiltless eye' (I. 333) synecdochically asserts that he is committing no trespass as he wanders beneath the remnants of 'fallen avenues' on an improved estate, and admires the chiaroscuro:

>   So sportive is the light
> Shot through the boughs, it dances as they dance,
> Shadow and sunshine intermingling quick,
> And darkening and enlightening, as the leaves
> Play wanton, every moment, every spot.
>   And now, with nerves new braced and spirits cheered,
> We tread the Wilderness, whose well-rolled walks,
> With curvature of slow and easy sweep –
> Deception innocent – give ample space
> To narrow bounds. The Grove receives us next.... (I. 345–54)

Although Cowper tends to use the same 'And now...' formulae as the topographical and picturesque poets, his verse is sufficiently infilled with sensuous detail to redeem the convention, and avert the appearance of an inventory of ready-made images. His unwillingness to provide a literary analogue to the school of paintings that 'throws Italian light on English walls' (I. 425) is authentic.

The network of positive and negative connotations of walking/wandering is more complicated in *The Task* than it is in most poets of the period, with social and religious contexts, and referential and symbolic language, overlapping in an unresolved tension. The social dimension of Cowper's ambivalence about walking is plainly evident in his encounter with a band of gipsies in Book 1: this 'vagabond and useless tribe' are thought to have abdicated their rational humanity in preferring 'squalid sloth to honourable toil', though he is forced to acknowledge the healthful aspects of 'breathing wholesome air, and wandering much', which produce a 'gaiety of heart' that presumably compares favourably with 'the gaiety of those / Whose headaches nail them to a noonday bed' treated earlier (I. 557–91, 499–500). Clearly Cowper's own appreciation of the freedom of rural walking comes into conflict with his suspicion of the social other and his Protestant disapproval of fecklessness and parasitism ('Loud when they beg, dumb only when they steal' [I. 573]).

The verb 'wander', in particular, is regularly overwritten in *The Task* with prejudicial religious significance. This is not invariably the case: 'Here unmolested, through whatever sign / The sun proceeds, I wander', Cowper writes in Book VI, going on to talk of his familiarity to the wildlife in his neighbourhood woods and fields, and of his derivation of a 'reasonable joy' from the imagined sight of 'animals enjoying life' (VI. 295–6, 347, 325). This is man performing his appointed role in the divine 'economy of Nature's realm' (VI. 579). But in the postlapsarian world wandering is also the readiest concretising metaphor for spiritual error: in Book V, for example, Cowper concludes a contemplation of the heavens with a comparison between the believer who has seen the lamp of truth, and who 'runs the road of wisdom', to him who 'wanders... bemazed in endless doubt' (V. 847–9); while in Book VI he rounds off a passage of millenarian speculation and an attack on Unitarianism with the grim reminder that 'All pastors are alike / To wandering sheep, resolved to follow none' (VI. 890–1). The more distant Cowper gets from the personal and quotidien in subject-matter, and the more seriously he engages an evangelical rhetoric, the more predictable it is that the idea of wandering becomes dogmatically opposed to directness as the way of truth. And yet he longs for the innocent freedoms of unfallen humanity, and his favourite image of such freedom is a humble locomotive one: in the rare moments of confidence in personal salvation one dares to claim 'To walk with God, to be divinely free' (V. 722), and at times in his wanderings around Olney Cowper finds that contiguity with nature inspires him with belief in contiguity with its creator:

> Happy who walks with him! whom what he finds
> Of flavour or of scent in fruit or flower,
> Or what he views of beautiful or grand
> In nature, from the broad majestic oak
> To the green blade that twinkles in the sun,
> Prompts with remembrance of a present God.   (VI. 247–52)

In concluding *The Task*, Cowper defends his choice of a modest, contemplative life, disdaining the pleasures of the world, and performing humble good offices: he casts himself as one who, like Isaac, 'Walks forth to meditate at eventide, / And think on her ['the World'], who thinks not for herself' (VI. 949–50). He also refers back to the origin of the *The Task* in a request for a poem on a Sofa – 'To dress a Sofa with the flowers of verse' (VI. 1007) – from a lady 'fond of blank verse', and in so doing comes close to linking the pedestrian context of his meditations with the itinerant progress of his text:

> I played awhile, obedient to the fair,
> With that light task; but soon, to please her more,

> Whom flowers alone I knew would little please,
> Let fall the unfinished wreath, and roved for fruit;
> Roved far, and gathered much.... (VI. 1008–12)

Cowper's 'Advertisement' also states that, having begun with the sofa as subject, he 'connected another subject with it; and pursuing the train of thought to which his situation and turn of mind led him, brought forth at length, instead of the trifle which he at first intended, a serious affair – a Volume'. It is at best a speculative set of relations, but it does seem irresistible to posit a synergy between Cowper's peripatetic habits, the loose, open-ended connectivity of *The Task* as improvised epic, and blank verse as the enabling medium of its languid philosophical travels.

Blank verse is indeed the poetic medium in which the first generation of writers who lived in, and helped construct, the 'great age of pedestrianism', embodied their experience to best effect, as I shall demonstrate in the following chapters. The middle classes were asserting new freedoms in travel and recreation at the same time as unrhymed pentameter (*versi sciolti da rima*, 'verse freed from rhyme', the Italian term, has a more positive connotation than 'blank verse') was acquiring a syntactic flexibility not seen since the Renaissance and a rhythmic freedom that allowed it to represent the speaking voice, or inner speech, with unparalleled assurance. The peculiarly fine adaptation of this rejuvenated blank verse to peripatetic themes is such as to lend dignity and authenticity to otherwise quite indifferent poetry.

As an example, take 'The Walk' by Miss M. Bowen, a woman poet based in the west of England who has eluded even the encyclopedic researches of Janet Todd into female literary history.[25] Bowen begins her poem with conventional exclamations at the sublimity of the Avon Gorge, homes in on the differently-coloured strata, then offers alternative religious and quasi-scientific speculations on the formation of the 'mighty chasm' (l. 27). The sight of a boat coming up the Avon to Bristol docks inspires reflections on the 'commercial interests' (l. 68) that dictated the construction of the towpath, which in turn prompts memories of flowers that used to grow there, and of her brother (now absent in the war) who used to pick them for her. There are vestiges in Bowen's verse of the formulaic rhetoric of the eighteenth-century topographical poem with its fondness for a punctuated series of contrasting vistas, but there is just enough momentum in the syntagmatic chain to recuperate these clichés and create a sense of the physical-perceptual reality of a walk:

> Now, as we lower trace the river's course,
> The prospect opens, we have left behind
> The lofty rocks and overhanging crags,
> And nothing now doth greet the ravish'd sight
> But graceful slopes and richly planted meads,
> And the smooth surface of the distant sea.
> ....
> Hence we ascend the steep with gradual step,
> Laborious task! but more than amply paid,
> By charms expansive rising on the sight:
> Still wider grows the clear translucent wave,
> Glowing with varied evening's lovliest tints... (ll. 109–14, 122–6)

A genuine ruined fortification occasions a reference to ancient battles against the Romans, who were offended by 'the boast / Of freeborn Britain' (ll. 256–7); inevitably, this is followed by the pious hope that the present war will end triumphantly. There is no sense – in contrast with *Cooper's Hill* – that topographical details have been fabricated to permit the expression of prepackaged sentiments. Bowen concludes by inviting her Muse to 'descend the steep' (l. 272) with her, and thanks her for assistance in describing 'whate'er / In nature pleas'd us as we've stray'd along' (ll. 276–7). In fact, as I have indicated, her poetic walk encompasses more than nature description: it is a variegated, unsynthesised record of observations, anecdotes, personal memories, and moral, historical, and political reflections, arranged along and motivated by the metonymical sequence of the walk. Referentially grounded or not, the walk is a potent trope for centering the representation of a subject, a perceiving and thinking self that achieves a minimum of stability in inverse proportion to the slowly altering surroundings. For all its awkwardnesses, Bowen's blank verse, like Cowper's, offers valuable glimpses of this subject-in-motion, an entity very different to the transcendental ego pilloried by critiques of Romantic ideology.

The question I have begged is why blank verse, beginning with writers like Crowe and Cowper and reaching an apogee in the poetry of Coleridge and Wordsworth, should have proved a particularly fit vehicle for peripatetic themes. It is, of course, noticeable that important prosodic terms such as 'foot', 'enjambement' and 'dipody' allude to the action or bodily means of walking, as do many of the impressionistic adjectives often applied to poetic rhythms, such as 'plodding', 'brisk' or 'pedestrian' itself. But these are perhaps no more than incidental signs of a more fundamental analogy between walking and poetry, of which creative writers themselves have not been unaware. The case of Wordsworth will be examined in the next chapter; A. R. Chisholm cites the more recent case of Paul Valéry:

> In his Oxford lecture, *Poésie et Penseé abstraite*, delivered in 1939, Valéry has told us of the relationship that he once noticed between the rhythm of his walking and the rhythm of an as yet unshaped poem. He thus seems to imply (1) that there is an almost physiological element in the making of poetry, and (2) that rhythm precedes words in the genesis of a poem.[26]

Both implications seem relevant to the production of blank verse peripatetic. To begin with, the claim that there is a physiological element in the making of poetry, at least so far as its rhythmic character is concerned, now appears to enjoy wide currency among prosodic theorists. David Abercrombie, building on the 'motor phonetics' of R.H. Stetson, explained long ago how the syllabic structure of language is the product of regular chest-pulses dividing the air stream prior to articulation, with stress the outcome of intermittent, more powerful muscular contractions. 'All rhythm, it seems likely', Abercrombie states, 'is ultimately rhythm of bodily movement', and poetic rhythm results from a particular organisation of the 'sound-producing movements' of speech.[27] On a similar basis, D.W. Harding has made an interesting study of the expressive potential of rhythm in poetry: when speech rhythm is innately a matter of the disposition of muscular energy, it is no surprise that rhythm can be exploited poetically to suggest other forms of movement or energy expenditure, among which Harding specifically names 'manner of walking' (though he is careful to add that the rhythm can only reinforce 'what the sense suggests').[28] He also supports the view, long maintained but often fancifully inflated, that rhythm may enhance the representation of states of mind or feeling, because of the correlation between rhythm and 'the energy conditions that accompany emotion'.[29] Harding's summary of his position on the mimetic functions of rhythm again underscores the analogy with walking among other muscular activities:

> The literary significance of rhythm and rhythms can best be understood by regarding language movement...as comparable to such systems of bodily movement as walking, gesture, and patterns of changing posture. Like these it can be described in terms of broad characteristics – flowing, jerky, patterned, disjointed, and so on – which give rise to similarly broad aesthetic appraisals, the total effect being in some degree pleasing or unpleasing, and described in such terms as graceful, charming, harmonious, clumsy, affected, unlovely.[30]

Prosodists, in fact, reach naturally for pedestrian comparisons when discussing the physical properties of rhythm, whether in speech or verse. Derek Attridge's lucid account of the rhythmic structure of English speech highlights its dual principles of the syllable and stress, and his discussion of each feature makes a suggestive link with pedestrian activity: the syllable, he writes, is 'the smallest *rhythmic* unit of the language', and 'like the step in walking, it is the repeatable event which keeps the utterance going'; the stress-timed movement of English (the occurrence of stresses at perceptually equal intervals) combines with the syllabic principle to produce a strong tendency towards the alternation of stressed and unstressed syllables. This has its origins in involuntary muscular processes, since 'we prefer to use our muscles in a rhythmic way for repeated actions, like breathing or walking, and we should not be surprised to find that a regular sequence of energy expenditure and relaxation forms the basis of our speech activities'. As Attridge points out, the alternating tendency is impurely realised in ordinary language use, but forms the basis of the structure of English verse, 'which capitalises on both the satisfying sense of regularity produced by bringing the two rhythmic tendencies into accord, and the expressive possibilities inherent in the conflict between them'.[31]

Attridge argues that the four-beat line is the most powerful underlying rhythm in English poetry, and that this is built up by a doubling principle into the 'underlying rhythmic structure' of the 4×4 stanza that is the basis of popular verse and song. This structure is characterised, as well as by its peculiarly insistent rhythm, by a strong propensity for rhyme and definite line-endings (that is, an identity of rhythmic and syntactic units). These features are absent from the five-beat line, which has a less urgent and perceptible underlying rhythm, but which by this very token, in the opinion of Attridge as of most other prosodists, is more responsive to the natural rhythm of spoken English. This determines other characteristics such as an overwhelming preference for duple rather than triple rhythms, a greater strictness in the syllabic count than in the four-beat line, and a resistance to the dipodic tendency (an alternation of primary and secondary stresses) found in popular verse. In addition, because the pentameter line has no predisposition to group into some larger rhythmic structure, each such line is an independent unit, and syntax decides whether one pauses or reads on at the line-end. It is this that encourages the absence of rhyme and the flexible run-ons that are the distinguishing marks of blank verse.

My suggestion, on the basis of the widely-shared understanding of the physiological roots of poetic rhythm that I have summarised, is that when, towards the end of the eighteenth century, the poetry-writing classes began to discover and express their bodily freedom in excursive walking or the more radically liberating pedestrian tour, they veered increasingly (and largely unwittingly) towards blank verse as the metrical structure best suited to *embodying* their experience and the mentality it engenders, because it was here that the muscular rhythms indigenous to both walking and poetry were brought into an intuitive correspondence. It remains a *felt* or *perceived* correspondence that does not entail the transparently absurd proposition that the movement of blank verse can evoke pedestrian activity in the absence of any relevant semantic indications from the words used. It is rather the case that the rhythmic qualities of blank verse – a steady alternating rhythm freely sustained across line-boundaries, with variable pauses within the line – can function iconically when the peripatetic theme is part of the signified meaning of the text. In the case of a poet like Wordsworth, for whom walking contributed one of the habitual rhythms of daily life, it may well be that rhythm preceded words in the genesis of poetry; but the mimesis of walking by the movement of language in blank verse is unlikely to take hold for the reader unless the link has been established by the normal processes of signification. Where such a link has been established, however, a powerful rhythmic association can be created: each syllable may be felt as a tread, with the underlying duple rhythm of stressed and unstressed syllables evoking the regular alternation of right foot and left foot,[32] the ongoingness of walking simulated by the 'striding-across' of enjambement, and with pauses determined not by force of metrical convention but, as they are in walking, by the need to take breath. The continuity and coherence of a significant amount of Romantic blank verse may therefore derive not from the immaterial unity of symbolic language, but, in a real sense, from a body in motion.

4
William Wordsworth: Pedestrian Poet

Wordsworth inevitably has a prominent place in a study of Romantic pedestrianism, because of the sheer volume of his poetry that involves, or is involved in, walking, and because of the heterogeneous nature and considerable aesthetic interest of those involvements. He was, as is well known, a prodigious walker: De Quincey reports that the poet's legs 'were pointedly condemned by all the female connoisseurs' that he had come across, but is obliged to point out that they were nonetheless very serviceable legs, having traversed, by his somewhat mysterious calculations, 'a distance of 175 to 180,000 English miles'.[1] Upwards of 2000 of these miles were comprised by the walking tour of 1790, in the opinion of its most painstaking chronicler.[2] As I have shown, this expedition, in which Wordsworth and his companion, Robert Jones, averaged about thirty miles a day, was by no means as exceptional as has sometimes been assumed, but it was still an impressive undertaking and establishes Wordsworth as one of the most formidable Romantic walkers.

However, it is the fluid and often highly mediated relationship between this physical pedestrianism – or, rather, a sensibility grounded in and shaped by regular and strenuous walking – and the poetic representation of walkers and walking in Wordsworth's early poetry, as well as what I will later posit aesthetically as a kind of textual pedestrianism, that I want to examine in this chapter. There is a range of means, all the way from polemical simplicities of theme to more speculative 'infrastructural' determinations, by which walking generates writing. Anne Wallace has made a persuasive case for Wordsworth as the originator of a genre she calls 'peripatetic', which she derives ultimately from Virgilian georgic,[3] and which I am happy to appropriate as the master-term for my own rather different theorisation of the link between pedestrianism and Wordsworth's poetic practice. The georgic parentage Wallace ascribes to Wordsworthian peripatetic steers her analysis heavily in the direction of *Home at Grasmere* and *The Excursion*, and she has surprisingly little to say about the series of longer poems with pedestrian themes or structures, up to and including *The Prelude*, that I intend to concentrate on. This is partly a question of different forms of pedestrianism: there is a distinction between the local, bounded, circular, 'excursive' walking highlighted by Wallace, and the more fluid, improvised, open-ended walking typified by the pedestrian tour – and these different types of walking generate different, and in some ways contradictory, aesthetic models. Wallace, in favouritising the pedestrian and aesthetic practices of the more mature, geographically rooted poet, has obscured a whole dimension of Wordsworthian peripatetic which arguably shows him at his most exciting and innovative.

One of my initial spurs to carrying out this research was a remark in an essay by Seamus Heaney that 'Wordsworth at his best, no less than at his worst, is a pedestrian poet.'[4] Heaney no doubt had his tongue wedged firmly in his cheek when he wrote this, but it is an insightful and suggestive remark. What he has in mind, most directly, are Wordsworth's compositional habits. His overwhelming preference for composing outdoors is well-known and well-documented: in order to write 'Michael', for example, he ascended day after day from Dove Cottage to a straggling heap of stones such as he describes at the beginning of the poem. He also had a marked tendency to compose peripatetically: in the blank verse meditation beginning 'When first I journeyed hither' (in the early version),[5] he writes of his delight in discovering that his brother John, whilst staying at Grasmere, had worn a path by 'habitual restlessness of foot' (l. 71) within a favourite nearby fir grove, where Wordsworth himself had previously found the trees planted too close together. Now, perhaps 'Timing [his] steps' (l. 113) to those of John, a 'silent Poet' (l. 88), on the deck of his ship, he can indulge his desire to 'walk / Backwards and forwards long as I had liking / In easy and mechanic thoughtlessness' (ll. 36–8). This circumscribed freedom of movement – a hypnotic, metronomic walking within a protective space – is what he found best loosened his creative faculty, as Hazlitt was later to describe in his classic essay, 'My First Acquaintance with Poets': whereas Coleridge 'liked to compose in walking over uneven ground, or breaking through the straggling branches of a copse-wood', Wordsworth preferred 'walking up and down a straight gravel-walk, or in some spot where the continuity of his verse met with no collateral interruption'.[6] For Heaney, the length of the gravel-walk must have corresponded with a line of blank verse, and as Wordsworth crunched up and down the path the regular physical motion accommodated the prolonged equilibriums of his style. Up-and-down walking, he says, does not forward a journey but habituates the body to a languorous, dreamy rhythm – a rhythm analogous to the slow, cumulative movement of Wordsworth's philosophic verse. This suggestion certainly offers an intriguing variation on the prosodic connection between walking and poetry that I considered in the last chapter. What Heaney says may well be true of Wordsworth's blank verse at a certain sub-verbal (or paralinguistic) level, but it far from exhausts the potential of the genetic link between walking and writing, and may be misleading if it gives the impression that a languorous, dreamy, ambulatory rhythm necessarily produces a similarly languorous and dreamy poetry. I shall be arguing instead that many of Wordsworth's early poems written during or about journeys undertaken on foot, however actually composed on the ground (as it were), take their very energy from 'collateral interruption', broadly understood.[7] If Heaney's remark possesses intuitive validity, we must nevertheless posit a regular divergence in Wordsworthian peripatetic between physical origins and discursive ends, between material base and linguistic superstructure: an up-and-down bodily/verbal rhythm may accommodate less easily regulated freedoms at the level of meaning. Textual pedestrianism may not take kindly to the discipline of the gravel-walk.

How, then, might one build on Heaney's interesting speculations and begin to unravel the complex truth of his assertion that Wordsworth *at his best* is a pedestrian poet? My answer to this question will involve surveying a range of Wordsworth's poetry that focuses on walking or the figure of the walker. In fact, this embraces all the important longer poems in the modern canon of the early Wordsworth: *An Evening Walk*, *Descriptive Sketches*, the Salisbury Plain poems, *The Borderers*, *The Ruined Cottage*, *Home at Grasmere* and significant parts of *The Prelude*. These cover the decade of Wordsworth's life – the 1790s – when his social and economic position was most insecure, as well as the first few years at Grasmere, when he was geographically more rooted, but still encumbered with worldly and vocational anxieties.

My discussion of the picturesque in Chapter 2 should provide a frame for considering the idiosyncrasies of *An Evening Walk*.

Written in 1788–89 and published in 1793, this poem has traditionally been regarded, not too favourably, as a close poetic cousin of the prose literature of picturesque tourism, and, insofar as it is a descriptive poem, as an elaboration of a repertoire of stylised elements from the topographical poetry tradition. R.A. Aubin, in his standard treatment of the genre, writes:

> Conventionally enough it opens with a statement of the poet's situation far from his friend and proceeds to the early memories theme, moralizing..., genre scenes, modesty and local pride..., humanitarianism, prospect, water-mirror, 'pensive, sadly-pleasing visions,' and retirement.[8]

Aubin's satirical tone comes from reading too much bad topographical poetry, and he does concede that Wordsworth conveys 'the sentiment of nocturnal mountain calm' better than any earlier poet, but his characterisation would not be disputed by most casual readers of the poem, certainly as far as its first two thirds is concerned. Wordsworth assembles sights and sounds from his native Lakeland with due respect for picturesque subjects, qualities and perspectives, and does so in the restrained, generalising way associated with picturesque art. The Fenwick note states that *An Evening Walk* was not 'confined to a particular walk or an individual place', out of an 'unwillingness to submit the poetic spirit to the chains of fact and real circumstance';[9] and, as J.R. Watson has pointed out, the opening eight lines, which offer a concatenation of images from Borrowdale, Derwentwater and Rydal, immediately confirm this trend to delocalised description.[10] The 'forest glooms', 'opening lakes' and 'tremulous cliffs' (ll. 4–6)[11] could easily be anywhere else. Neither, correspondingly, is there much sense of the ongoing temporality of a walk: as Alan Liu has observed, the poem's verse paragraphs tend to group descriptive details into a 'simultaneous tapestry view', or a number of such views;[12] and when there is a shift in time, as in the transition from noon to evening in lines 85–8, it appears curiously as more of a spatial turn, as though one could swivel one's eyes from midday to twilight. Instead of a simulated pedestrian perspective, therefore, Wordsworth largely honours the preference for fixed viewpoints or 'stations' that facilitates the picturesque gaze, and so to read *An Evening Walk* as peripatetic one has to read the several displacements which walking undergoes in the text.

To begin with, there is the sense in which the arrested motion of the pedestrian is projected onto the outside world: the group of potters 'Winding from side to side up the steep road' (l. 110), the peasant launching himself on his sledge down the 'headlong pathway' (l. 112), the cockerel 'Sweetly ferocious round his native walks' (l. 129), the quarry-workers in the distance who 'O'erwalk the viewless plank from side to side' (l. 148), the 'violent speed' (l. 180) of the horsemen shadows on the twilight hills, all partake of the physical animation which the walker, rooted in a succession of 'stations', is not permitted to enjoy. More subtly, there is the sense in which the eye wanders among the intricacies of the poem's descriptions. As I argued in Chapter 2, the freely-deviating line of the pedestrian's itinerary is a mirror-image of the variety and intricacy of picturesque nature, whose eroticised appeal is predicated on perpetual partial concealment. One definition of textual pedestrianism is therefore that which stresses the insistence of half-glimpse, the fleeting penetration and withdrawal of light into and out of shade, as in the description of sunset in lines 155–64:

> And now it touches on the purple steep
> That flings his shadow on the pictur'd deep.
> Cross the calm lake's blue shades the cliffs aspire,
> With tow'rs and woods a 'prospect all on fire;'
> The coves and secret hollows thro' a ray
> Of fainter gold a purple gleam betray;
> The gilded turf arrays in richer green
> Each speck of lawn the broken rocks between;
> Deep yellow beams the scatter'd boles illume,
> Far in the level forest's central gloom....

Here the pedestrian is frozen in contemplation, but the eye (of the text) is made to wander instead, seduced by the swiftly-altering view of mountain shadows, 'blue shades', coves, 'secret hollows' and gloomy forest that await momentary illumination. Kim Taplin writes of how imagery of 'untrodden ways' – illustrated here by secret hollows and the central gloom of the forest – often suggests a female sexuality barred to men: 'Footpaths', she says, 'are our routes to a licensed intimacy with the landscape, to a carnal knowledge of nature.'[13] In *An Evening Walk*, where there is no consummation, but no end of stimulation, the path to such intimacy is chiefly a visual one: it is a sexually errant eye that performs the hesitant intrusion. The picturesque gaze is full of hungry movement.

In the final third of the poem this gaze is ruffled decisively by the introduction of the Female Beggar. It has been noticed by many commentators that this section, together with the description of the swans that precedes and motivates it, is anomalous in dwelling at length on a single object in the landscape, in contrast to the poem's accustomed mode of quickfire accumulation of balanced, opposing images. But it has seldom been noticed that in turning to the Female Beggar the pedestrian observer is expressly turning *away from* observation: in spite of the fact that Wordsworth uses a language of perception ('I see her now'), it is clear beyond doubt that the Beggar is an imaginary figure, the pure product of the narrator's tormented fantasy rather than the nominal object of an encounter on either an actual or an idealised walk.[14] This is where we enter upon fully-internalised walking, which paradoxically sweeps away the perfunctory rhetoricity of the peripatetic motif and gives us walking as real, strenuous, prolonged physical motion:

> Fair swan! by all a mother's joy caress'd,
> Haply some wretch has ey'd, and call'd thee bless'd;
> Who faint, and beat by summer's breathless ray,
> Hath dragg'd her babes along this weary way;
> While arrowy fire extorting feverish groans
> Shot stinging through her stark o'erlabour'd bones.
> – With backward gaze, lock'd joints, and step of pain,
> Her seat scarce left, she strives, alas! in vain,
> To teach their limbs along the burning road
> A few short steps to totter with their load....   (ll. 242–50)

Wordsworth follows the woman through the forest, where her eldest child asks questions that betray his ignorance of his father's death, along the 'painful road' (l. 271), where she amuses the children with glow-worms, to the fatal conclusion of her journey on the 'lightless heath' (l. 285), where we leave her with the two babies 'coffin'd' (l. 300) in her arms. In the course of this journey, pedestrian travel is allegorised as the endurance of Hope:

> And bids her soldier come her woes to share,
> Asleep on Bunker's charnel hill afar;

> For hope's deserted well why wistful look?
> Chok'd is the pathway, and the pitcher broke.   (ll. 253–6)

Walking is synonymous with an instinct for life, while the choked pathway with which the poet obstructs the Beggar's stubborn last hopes suggests the threat of engulfment that overtakes his own pleasurable participation in the natural scene. On returning to the poem's descriptive mode with a night-piece that mixes aural with visual stimuli to convey the 'restless magic' (l. 345) of nature under a different aspect, the psychologised significance of walking which the Female Beggar passage has established lingers on. There is an element of crisis-management in this final section of the poem, the fall of night occasioning anxieties of introversion for the walker whose thoughts have been taken up for too long by imaginary horrors and who requires an outer scene in which he can see his broodings re-enacted in a more manageable (because aesthetically ordered) 'pageant scene':

> Unheeded Night has overcome the vales,
> On the dark earth the baffl'd vision fails,
> If peep between the clouds a star on high,
> There turn for glad repose the weary eye;
> ....
> Nought else of man or life remains behind
> To call from other worlds the wilder'd mind.   (ll. 363–6, 375–6)

The 'other worlds' are the inner terrors which a continuation of the walk, and the world of distracting phenomena it represents, might help to dissipate. But the poet is still mentally wandering and suffering with the beggar whose plight he has imagined, as verbal echoes help to reinforce: the woman who 'wilders o'er the lightless heath' (l. 285) is recalled in the night-duck who ceases clamouring for his 'wilder'd mate' (l. 357) to attend to the ephemeral beauties of twilight, and is then brought into more explicit alignment with the 'wilder'd mind' of the poet himself.

At this point the moon rises and crisis is averted – not, however, by reopening the nocturnal walker's field of vision, but by inspiring a symbolic prospect equating the onward journey with hope for the future. Hope is invested specifically in a rural retreat to be shared with his sister:

> Thus Hope, first pouring from her blessed horn
> Her dawn, far lovelier than the Moon's own morn;
> 'Till higher mounted, strives in vain to chear
> The weary hills, impervious, black'ning near;
> – Yet does she still, undaunted, throw the while
> On darling spots remote her tempting smile.
> – Ev'n now she decks for me a distant scene,
> (For dark and broad the gulph of time between)
> Gilding that cottage with her fondest ray,
> (Sole bourn, sole wish, sole object of my way...)   (ll. 407–16)

Only with the thought of this future idyllic retirement is Wordsworth able to restore, in the final verse paragraphs, a picture of a harmonious natural scene, though it is an unquiet night with which he concludes his 'walk', and the concluding four lines in particular – with the sob of the owl, the mill-dog's howl, the thump of the forge and the yell of the lonely hound – give a nervous edge to the tranquillity. By way of resolving the problem caused by the poet's imaginative and rhetorical entanglement with a vagrant woman, his own onward journey becomes an allegory of hope and desire, in what will become a familiar Wordsworthian revision of a very traditional poetic topos.

The poet and the Beggar therefore present positive and negative inflections, in different ratios of literal and metaphorical, of the idea of life-as-walking. Neither has much to do with the walk which forms the ostensible subject-matter of the poem, and this serves notice that Wordsworth's peripatetic art will be marked by detours and displacements. We began by seeing how the overcoded proprieties of the picturesque gaze are discomfited by the energies conspicously suppressed in that gaze or projected into the natural world. The Female Beggar episode, however, is responsible for the most glaring of the several faultlines in the poem – discontinuities or 'sudden variations' of style and subject-matter that produce an intriguing aesthetic 'roughness', suggesting that a poem which begins in control of the picturesque code has become performatively contaminated by that same picturesqueness. It is in this sense that *An Evening Walk* helps us see the potential in Wordsworth for a *textual* pedestrianism, an itinerant poetic mode that resists argumentative or visionary closure just as it resists the complacent integrity of a poem fully at ease in its genre.

Descriptive Sketches, which was published simultaneously with *An Evening Walk*, is based on Wordsworth's walking tour of 1790, but was written in 1792 in a state of more informed enthusiasm for the French Revolution and under the strains of his relationship with the now pregnant Annette Vallon. Its mood of studied melancholy has been ridiculed by critics as inconsistent with the upbeat mood recorded in the only letter written on the tour to have survived: Wordsworth writes there that his spirits 'have been kept in a perpetual hurry of delight by the almost uninterrupted succession of sublime and beautiful objects which have passed before my eyes during the course of the last month',[15] and the only sadness he acknowledges stems from the thought of leaving the magnificent Swiss scenery, not from the 'wounded heart' invoked in the opening lines of *Descriptive Sketches*.[16] Clearly the poem's ostentatious mood of dejection is the retroactive product of the time of composition, when it would make more sense for Wordsworth to portray himself as one who 'plods' through nature seeking her 'varying charms' to distract him from the 'sad stroke of Crazing Care / Or desperate Love', and to make teasing allusion to 'A heart, that could not much itself approve' (ll. 15–6, 43–4, 46).

The dedication to Robert Jones prefaced to the 1793 edition of *Descriptive Sketches* shows Wordsworth's awareness of the gap 'between two companions lolling in a post chaise, and two travellers plodding slowly along the road, side by side, each with his little knapsack of necessaries upon his shoulders', grounding the distinction in the affective bond between the pedestrian travellers. If there is an element of proud humility in this sentimental characterisation of pedestrian travel, it perhaps links with the poet's political radicalism, most plainly expressed in the violent imagery announcing the dawn of Liberty that concludes the poem. This radicalism again belongs most obviously to the period of composition and publication, which is roughly contemporary with Wordsworth's republican *Letter to the Bishop of Llandaff*, rather than to the period of the walking tour itself: the letter to Dorothy of September 1790 remarks that France (which receives only a handful of lines in *Descriptive Sketches*) is 'mad with joy, in consequence of the revolution', but can only add blandly that 'we had many delightful scenes where the interest of the picture was owing solely to this cause'.[17] This latter sentence provides a revealing gloss on the state of Wordsworth's aesthetic, as well as political, education at the time of the tour. His fidelity to the norms of picturesque tourism and the codes of picturesque beauty is such that he can appraise a people's revolutionary fervour only in terms of a novel foreground to pictorial composition. And despite his reference to a 'perpetual hurry of delight', he seems only too well schooled in the practice of the picturesque gaze:

> Ten thousand times in the course of this tour have I regretted the inability of my memory to retain a more strong impression of the beautiful forms before me, and again and again in quitting a fortunate station have I returned to it with the most eager avidity, with the hope of bearing away a more lively picture.[18]

However, by the time he came to translate those 'lively pictures' into the couplets of *Descriptive Sketches*, Wordsworth has, in his own opinion, outgrown the limitations of the picturesque: his note to line 347 claims that the Alps would be insulted by being so described, and, with a nod at the importance attached in picturesque art to contrasts of light and shade, he points out that if he had wanted to 'make a picture' of his stormy sunset he would have 'thrown much less light into it'. Instead he respected the emotional power of the scene by casting it in the stronger, more impassioned language of the sublime. Reinterpreting this remark, Gerald Izenberg has argued recently that it is Wordsworth's appropriation of the power of the natural sublime for the purposes of political struggle or imaginative self-aggrandisement that is the 'central aesthetic event' of *Descriptive Sketches*.[19]

If Wordsworth is concerned to show that he is no longer the picturesque tourist he was (in part, at least) at the time of the walking tour itself, he nevertheless cannot escape entirely from the structures of tourist experience, for all the difference which his pedestrian mode of travel imparts. His opening lines on the melancholy foot-traveller make an effort to blend the latter in with local people and local activities, but the very act of denying distance between the traveller and the indigenous people of the host country paradoxically works to reinstate it:

> Kind Nature's charities his steps attend,
> In every babbling brook he finds a friend …
> Host of his welcome inn, the noon-tide bow'r,
> To his spare meal he calls the passing poor …
> With bashful fear no cottage children steal

> From him, a brother at the cottage meal,
> His humble looks no shy restraint impart,
> Around him plays at will the virgin heart.
> While unsuspended wheels the village dance,
> The maidens eye him with inquiring glance....
> (ll. 27–8, 31–2, 37–42)

Whereas the traveller practises charity to the passing poor, Mother Nature provides his own emotional and spiritual handouts. Local people are said to treat him unaffectedly as a 'brother', but the insistent negative constructions nevertheless leave a trace of the 'bashful fear' and 'shy restraint' which they deny. These tensions are present throughout the poem. 'Delicious' scenes, such as the small organic communities united by the call to mass on the shores of Lake Como, are greeted by the eye only to receive an equally prompt farewell as the poet continues 'With pensive step to measure my slow way' (l. 165). In the more mountainous region of Underwalden, where 'no trace of man...profanes' the deep tranquillity, Wordsworth is again conspicuously a passing spectator of the rural idyll: the only dissonant sound is a shepherd-boy shouting to 'the stranger seen below' (ll. 425, 440). In this still early phase of the British tourist invasion of the Continent, Wordsworth's *Sketches* outline the contradiction between the traveller's self-consciousness as a transitory observer of foreign people and places, and his/her desire to make meaningful contact with the realm of novelty and difference that is the object of the journey, that has marked tourist experience ever since.

A related contradiction emerges between an emphasis on human fellowship and community values and the strong images of solitary dignity and endurance. Wordsworth's sympathy for the Grison gypsy seems deepened by the fact that she is excluded from the comfort and support to be found in shared suffering:

> – The mind condemn'd, without reprieve, to go
> O'er life's long deserts with it's charge of woe,
> With sad congratulation joins the train,
> Where beasts and men together o'er the plain
> Move on, – a mighty caravan of pain;
> Hope, strength, and courage, social suffering brings,
> Freshining the waste of sand with shades and springs. (ll. 192–8)

But the section adverting to the tradition of the Golden Age of the Alps, and to the ancient Swiss as a natural republican possessed of liberty and an instinctive morality, brings in its train a clear identification with his descendant, the modern-day mountain-dweller, similarly 'free, alone and wild':

> Ev'n so, by vestal Nature guarded, here
> The traces of primaeval Man appear.
> The native dignity no forms debase,
> The eye sublime, and surly lion-grace.
> The slave of none, of beasts alone the lord,
> He marches with his flute, his book, and sword.... (ll. 528–33)

Here, as in the lines where Wordsworth pictures himself standing alone gazing admiringly at the 'fearless step' of the distant chamois-chaser (ll. 366–79), independent fortitude in the face of a harsh and unmerciful nature takes precedence over the anonymous solidarity of a 'mighty caravan of pain'. There is clearly a strong emotional charge attached in *Descriptive Sketches* to scenes of settled domesticity in small mountain villages, but this is counteracted by an element of elective solitude and individual freedom that helps make the role of tourist a far from inappropriate vehicle for the poet's sensibility.

One further element in *Descriptive Sketches* worth noting is the inclusion of a peripatetic version of the life-as-journey metaphor similar to that in *An Evening Walk*, and which functions as a bridge between a description of the 'homely pleasures' (l. 582) of the mountain shepherd and a warning of the disasters – both natural, in the form of avalanches, and human, in the apparently inexorable separation of father and son – that are visited upon him by the *genius loci*:

> – Alas! in every clime a flying ray
> Is all we have to chear our wintry way,
> Condemn'd, in mists and tempests ever rife,
> To pant slow up the endless Alp of life. (ll. 590–3)

These lines are an example of the occasional 'outbursts couched in the language of an eternal human condition' which Izenberg sees as discrepant with the poem's dominant tendency to read human suffering as the effect of political oppression.[20] This is not the only reason for the anomalous impression they create: although Wordsworth's poetry in the 1790s is densely populated by poor, destitute, oppressed and vagrant persons who might be appropriately figured as panting up the Alp of life, the latter is not a good characterisation of the pedestrian tourist in this poem, whose troubled journey is marked, sometimes very discreetly but still legibly, by a force of desire variously oriented towards sexual passion (of which the lines on 'Those shadowy breasts in love's soft light arrayed' [l. 154] are an uncommon and striking instance in Wordsworth), natural scenery and revolutionary politics, not always distinguishably. Repeatedly frustrated or interrupted by the passage of the text, this desire finally erupts in the apocalyptic vision of the triumph of Freedom at the poem's conclusion ('Oh give, great God, to Freedom's waves to ride / Sublime o'er Conquest, Avarice, and Pride ...' [ll. 792–3]), which puts paid to whatever is left of the serene attentiveness of the picturesque gaze as surely as the Female Beggar in *An Evening Walk*. In the development of Wordsworthian peripatetic, *Descriptive Sketches* shows clearly that although there is a strong inclination to dramatise the walker as an emblem of human hardship and suffering, in a way that respects the material history of walking as a largely involuntary, lower-class form of travel, there is also a determined poetic initiative by which walking becomes the physical correlate of the mobility of desire, and, in the context of a poem narrating a pedestrian tour, a metonym for a kind of textuality marked by the turn and counterturn of aesthetic adventure.

In the Salisbury Plain poems, it is the representation of a suffering female vagrant which again absorbs most immediate attention. These poems, which have a typically convoluted (for Wordsworth) textual history, have a loose biographical foundation in an enforced walk he took across the Plain in the summer of 1793: the carriage in which he and William Calvert had begun a tour of the west country was dragged into a ditch, Calvert promptly departed on horseback, and Wordsworth's 'firm Friends, a pair of stout legs, supported him from Salisbury, through South into North Wales, where he is now quietly sitting down in the Vale of Clwyd' (as Dorothy wrote at the end of August).[21] Wordsworth later reconstructed the events of the walk across the Plain in Book XII of the 1805 *Prelude*.

In the first version of the sequence, *Salisbury Plain*, we find two involuntary and hard-pressed walkers: a male traveller, who has suffered some turn of chance and reflects bitterly on the privileged rich, and who now measures each 'painful step' (l. 39)[22] with a sigh across the Plain, and a female vagrant, who is the victim of real economic oppression, she and her father having been persecuted and dispossessed by a local landowner. In this further variation on the life-as-pedestrian-journey theme, Salisbury Plain effects a sublime voiding of the pedestrian's itinerary in the seeming infinity of a vast terrain vacant except for 'wastes of corn that stretched without a bound' (l. 44). In this rawest of protest poems, furthermore, walking is a survivor's reflex action, the physical expression of enforced exile from the kind of happy domesticity glimpsed in isolated cottages on the Plain:

> Long had each slope he mounted seemed to hide
> Some cottage whither his tired feet might turn,
> But now, all hope resigned, in tears he eyed
> The crows in blackening eddies homeward borne,
> Then sought, in vain, a shepherd's lowly thorn
> Or hovel from the storm to shield his head.
> On as he passed more wild and more forlorn
> And vacant the huge plain around him spread;
> Ah me! the wet cold ground must be his only bed.   (ll. 55–63)

The traveller's unexplained penury and alienation from society are accentuated not only by allusion to scattered cottages on the Plain (such as the one described in lines 406–14) and the minimal comforts of the shepherd's hovel, but also by mention of the crows returning to their roosts. The persistent and deliberate contrast between an unavailable and heavily sentimentalised domesticity and the peregrinations of the homeless traveller repeats a pattern from *Descriptive Sketches*, though there is nothing in the later poem which explicitly mobilises the positive energies of the peripatetic instinct.

Salisbury Plain is more than a well-dramatised social protest poem, however, and it is by virtue of its 'surplus' meanings that it assumes the character of peripatetic in the more formal aesthetic sense I am trying to develop. In the *Prelude* account of the walk across Salisbury Plain Wordsworth highlights the hallucinatory moment in which he saw ancient Britons 'stride across the

Wold'[23] rattling their spears, then the ritual sacrifice by fire of living men in large wicker cages. These elements are less prominent but nevertheless vividly present in the earlier version, though they are placed at several removes as the experience of a 'swain' as reported by an old man whom the female vagrant has encountered (!), and are treated with a Gothic intensity that is later toned down in *The Prelude*:

> It is the sacrificial altar fed
> With living men. How deep it groans – the dead
> Thrilled in their yawning tombs their helms uprear;
> The sword that slept beneath the warriour's head
> Thunders in fiery air: red arms appear
> Uplifted thro' the gloom and shake the rattling spear.   (ll. 184–9)

Within the loose narrative structure of *Salisbury Plain*, this grisly vision of the cruelty and violence of prehistoric culture articulates with the powerful sentiments on the dehumanising effect of modern war placed in the mouth of the Female Vagrant:

> Better before proud Fortune's sumptuous car
> Obvious our dying bodies to obtrude,
> Than dog-like wading at the heels of War
> Protract a cursed existence with the brood
> That lap, their very nourishment, their brother's blood.   (ll. 311–15)

Stanza 48 recalls the imagery of ancient warfare and ritual sacrifice, explicitly juxtaposing them with sonorous denunciations of the more lingering miseries created by social injustice. This then leads into the confused (the confusion not helped by a number of missing lines) polemical conclusion denying that 'Truth with human blood can feed her torch' (l. 516), but calling with millenarian fervour on the 'Heroes of Truth' to accelerate the establishment of a new world with what seems a comparable lust of destruction:

> High o'er the towers of Pride undaunted rear
> Resistless in your might the herculean mace
> Of Reason; let foul Error's monster race
> Dragged from their dens start at the light with pain
> And die....   (ll. 543–7)

In this narrative and meditative passage from the cruelty and violence of primitive culture, to the oppression of modern war, to the future reign of Truth and Reason, walking – to take walking as the dominant figure of textual passage – may be seen as loosening temporal fixities, effecting connections between past and present, and between self and world, but leading to no final resolution of the accumulated tension: the 'terrors of our way' remain. *Salisbury Plain* as a *performative* walking poem therefore does something rather different to what it does on the surface level as a poem *about* two walkers across a bare landscape: it brings inner and outer into fluid interaction, freely combining the materials of observation with the materials of learning and memory and the materials of invention, with no overruling imperative for synthesis and closure.

If the textual pedestrianism posited here, with its ability to provoke thought and to respect the partial vistas of life-as-it-lived, is not inappropriately applied to *Salisbury Plain*, it is more unequivocally present in *The Prelude*, where as readers we are critically habituated to dividing our attention between the autobiographical life's-story and the 'eye and progress' of the song itself. The Salisbury Plain episode in Book XII of the 1805 text comes as the climax to a Book that has a strong pedestrian sub-plot to its central theme of the restoration of imagination. Wordsworth describes how he turned to the 'Pathways' and 'lonely Roads' (XII. 124) of common life, after his mental rehabilitation, in an information-gathering, truth-seeking venture that resulted in *Lyrical Ballads*. In fact, two complementary grounds are adduced for his long-standing love of public roads: one is his discovery that the roads are 'schools' (XII. 164) in which his hopes of the simple virtues to be found among uneducated people are fulfilled; the other is a more romantic and intuitive appeal:

> I love a public road: few sights there are
> That please me more; such object hath had power
> O'er my imagination since the dawn
> Of childhood, when its disappearing line,
> Seen daily afar off, on one bare steep
> Beyond the limits which my feet had trod
> Was like a guide into eternity,
> At least to things unknown and without bound.   (XII. 145–52)

The first explanation of his fascination is of a kind of 'beautiful' recognition, wherein things present themselves as perfectly in line with his prefigurings; the other a sublime invitation to the mind to overreach its capacities. Admittedly, in Book XII, as Theresa Kelley has pointed out, the sublime is brought in 'under the aegis of the beautiful',[24] and Wordsworth moves quickly to shutter the sublime prospects he opens up, but they *are* opened up and make a more determined appearance in the Salisbury Plain section.

Wordsworth leads in to the retelling of his reverie with positive sentiments on the fellowship of the living and the dead, of poets as connected evolutionarily in a 'mighty scheme of truth', and notes his hope that

> a work of mine,
> Proceeding from the depth of untaught things,
> Enduring and creative, might become
> A power like one of Nature's.   (XII. 309–12)

These thoughts are then professedly illustrated, in no very perspicuous manner, by the account of his wanderings on the 'bare white roads' (XII. 316) of Salisbury Plain, where solitude disorientated and deranged the walker into a dark and disturbing vision of ancient combat and acts of sacrifice. If this is Wordsworth's example of standing 'By Nature's side among the men of old' (XII. 297), what does it mean that he is brought by Nature into communion with primitive Britons and their culture of 'barbaric majesty' (XII. 326), that he participates imaginatively in sacrificial rites, whether as judge or victim? Such questions have certainly exercised a new generation of critics of *The Prelude*, though I wish here only to note their insistence. The dark reverie is followed by a gentler 'antiquarian's dream' (XII. 348), in which the mounds and circles on the Plain are interpreted as the work of Druid-astronomers, the means by which they expressed 'Their knowledge of the heavens' (XII. 346) – a knowledge not only fitted to the external world, in the familiar Wordsworthian terms, but a knowledge inscribed physically *on* that world. The vision ends serenely:

> I saw the bearded Teachers, with white wands
> Uplifted, pointing to the starry sky
> Alternately, and Plain below, while breath

Of music seem'd to guide them, and the Waste
Was chear'd with stillness and a pleasant sound. (XII. 349–53)

The substitution of this gentler visitation for the horrors of the foregoing lines is evidently meant to reprise the theme of the 'twofold influence' of nature ('From nature does emotion come, and moods / Of calmness equally') stated at the start of Book XII, though it may give credence to Kelley's argument about the subordination of the sublime to the beautiful at this stage of the poem's argument. It also provides an esoteric embodiment of Wordsworth's hope that a work of his might be assimilated to nature in its operations on the receptive mind, and helps to reassure him of his ability to maintain an 'ennobling interchange' between the objects of perception and the 'higher power' of imagination.

Whether or not this philosophical conclusion is felt to be an adequate gloss on the hallucinatory intensity of the Salisbury Plain passage, it is interesting that a very different set of questions arises in this retelling of the story to that which presented itself in *Salisbury Plain* – differences that result both from its very recontextualisation within an autobiographical epic, and from the altered way in which, as Wordsworth acknowledges, things 'may be view'd / Or fancied, in the obscurities of time' (XII. 354–5). My point here is that it is the uniquely Wordsworthian peripatetic poetic 'spirit' that is driving these transformations: in the itinerant progress of his poetry, Wordsworth arrives at a re-envisioning and reinterpretation of his experience on Salisbury Plain, and this provides for another temporary stopping-place, or philosophical 'viewpoint', in what is an always ongoing walk. In *The Prelude*, a poem in which literal walks invariably modulate into metaphorical journeys, and those journeys are figurally intertwined with the passage of the text, this itinerant quality to Wordsworth's argument is readily appreciated, but it is really more the condition of his entire imaginative life, stretching back to the Salisbury Plain poems and beyond.

Reverting, then, to the chronological sequence, it is worth observing that the narrative and rhetorical device whereby walking, as literal sign of the displacement caused by poverty, war, or other empirical causes, is blurred with its figurative role in rehabilitating the life-as-journey metaphor, is made to work even harder in the next redaction of the Salisbury Plain poem, *Adventures on*

Salisbury Plain, written between 1795 and 1799. In this version, which has a larger cast of minor characters, the Plain is the site of multiple journeys, all involuntary and laborious, with guilt now added to fear and the survival-instinct as a driving-force. In particular, the male traveller is more solidly characterised as a murderer on the run: having been press-ganged into serving in the American Revolutionary Wars, he has his claim for financial compensation rejected by the 'slaves of Office' (l. 91), and in a mood of anger and resentment kills a fellow-traveller on his journey home. This therefore introduces a new element of *criminal* vagrancy into the poem, and the sublime landscape and Gothic intrusions are made to function as a symbolic commentary on the traveller's guilt and despair: the sight of a gibbeted human body, for example, brings on a hallucination of boulders rolling across the Plain 'as if to sweep him from the day' (l. 123), and the reference to the wicker sacrifices of ancient times now stands in allusive relation to this and hence to the traveller's inner torment. At the end of the poem, after the traveller's guilt has been discovered and he has been hanged for his crime, his own body is gibbeted as an admonition to passers-by, and this bleak finale strangles any slight hope that might have been entertained for the onward journeys of any of the poem's other itinerant poor.

One especially interesting addition in *Adventures on Salisbury Plain* is a largely positive depiction of the communal way of life of a group of gipsies, who give assistance to the female vagrant, and who offer a seemingly impossible fantasy of 'vagrant ease' (l. 507) to which the downfallen respectable classes cannot aspire:

Semblance, with straw and panniered ass, they made
Of potters wandering on from door to door:
But life of happier sort to me pourtrayed,
And other joys my fancy to allure;
The bag-pipe dinning on the midnight moor
In barn uplighted, and companions boon
Well met from far with revelry secure,
In depth of forest glade, when jocund June
Rolled fast along the sky his warm and genial moon. (ll. 514–22)

The petty criminality that forms the substratum of the gipsies' way of life deter the female vagrant from joining their group, but the impression of happy communality within a carefree peripatetic existence is not repudiated. David Simpson has pointed out that gipsies were perceived as threatening by late-eighteenth-century society 'because they seemed to have *chosen*, rather than to have been exiled to a life of wandering and displacement', and are 'a counterimage to the central values of the majority society, those of labour and property'.[25] Wordsworth's ambivalence about the 'wild brood' in *Adventures on Salisbury Plain* offers us particularly clear early evidence both of what Simpson describes as the 'gypsy in his soul', and of the more conservative alter-ego which favours physical and mental stability.

The walking-as-exile motif modulates even more surely into walking-as-self-exile in Wordsworth's verse drama, *The Borderers*, written in 1796/97, the entire action of which consists of assorted destitutes and outlaws criss-crossing a blasted heath on foot. Walking is again made inseparable from indigence and suffering, especially in the case of yet another female beggar, introduced in Act 1 Scene 3, who has been bribed into becoming an instrument of Rivers's plot to seduce Mortimer into crime. It is, more arrestingly, a minimal human activity in a situation where normal morality and social constraints are suspended, an abatement of anxiety over the lack of spiritual anchors, and it is therefore the appropriate form of the hero's existential anguish: at the end of the play, Mortimer, under the full horror of the discovery of his error in bringing about the death of Margaret's father, projects for himself a kind of walking compulsion that will take him outside human society and all companionship and communication:

I will go forth a wanderer on the earth,
A shadowy thing, and as I wander on
No human ear shall ever hear my voice,
No human dwelling ever give me food
Or sleep or rest, all the uncertain way
Shall be as darkness to me, as a waste
Unnamed by man! and I will wander on
Living by mere intensity of thought,
A thing by pain and thought compelled to live .... (V.iii.265–73)[26]

In this prefiguration of a locomotion fuelled by a morbid remorse that, as Wordsworth's prefatory essay suggests in discussing the character of Rivers, will require constant provocation ('all his pleasures are prospective, he is perpetually ch[a]sing a phantom, he commits new crimes to drive away the memory of the past'), Mortimer is examplary of a host of figures in Wordsworth's early poetry who are, to paraphrase the line just quoted, by pain or thought compelled to *walk*. It is notable too that Mortimer refers to wandering across 'a waste / Unnamed by man': walking as (self-)exile takes place in a landscape that cannot be humanised by the naming-function of language. Refusing the sense of worldly belonging that comes with the assimilation of one's surroundings to language, Mortimer sentences himself to the sublime segregation of his own 'intensity of thought' and all that exists outside it and apart from it.

Little distance in time separates the writing of *The Ruined Cottage* from that of *The Borderers*, but it is usually taken that in the former Wordsworth's meditative blank-verse style (which I have speculated is the intuitive correspondent form of peripatetic) is wielded for the first time with full confidence, and his mature philosophy of nature finds its earliest convincing expression. From my own point of view, *The Ruined Cottage* is certainly a central and fascinating achievement of Wordsworthian peripatetic, offering an almost schematic treatment of the mixed moral economy embracing vagrancy and wanderlust, on the one hand, and domesticity and retirement, on the other, in Wordsworth's mature writing.

The aesthetic predilections of this poem are adumbrated in the opening twenty-five lines, which draw an interesting contrast between the anonymous narrator and a 'dreaming man' who is clearly a type of the picturesque tourist. The latter 'Extends his careless limbs beside the root / Of some huge oak' (oaks were the classic picturesque tree, with huge twisted roots being considered quintessentially picturesque), looking at a 'soft and distant' view of hills within the frame of its 'impending branches' (ll. 9–17).[27] Disowning this posture of contemplative sedation, whereby the natural scene is soft and soothing precisely by virtue of its actual and quasi-pictorial distancing from the observer, the narrator characterises his own relation to the physical environment as one of discomforting propinquity:

> Across a bare wide Common I had toiled
> With languid feet which by the slippery ground
> Were baffled still; and when I sought repose
> On the brown earth my limbs from very heat

Could find no rest nor my weak arm disperse
The insect host which gathered round my face
And joined their murmurs to the tedious noise
Of seeds of bursting gorse which crackled round.   (ll. 18–25)

If this is a rejection of the easily-gained stationary perspectives of the picturesque tourist in favour of the pedestrian traveller's un-idealised responses to a nature indifferent to his well-being or pleasure, it is also anti-picturesque with respect to the subject matter of the story that follows. The ruined cottage would be an exemplary subject for picturesque poetry or the foreground of a picturesque painting, but as such would be prized for its visual qualities and would be divorced from narrative time and from the operation of human sympathy or social concern. In the eighteenth century it was not unknown for landowners to employ fake 'hermits' to sit outside carefully-constructed hovels to give the coup de grâce to their picturesque landscape gardens; but the old man whom the narrator of Wordsworth's poem sees 'Stretched on a bench' (a very picturesque bench, edged with moss and 'studded o'er with fungus flowers') beside the ruined cottage (ll. 37–8) has a story to tell, and it is a story which rebukes any latent tendency towards a dispassionate connoisseurship of crumbling masonry and creeping vegetation. I have already quoted Gilpin's observation that moral and picturesque ideas do not always coincide,[28] and *The Ruined Cottage* seems determined from the outset to demonstrate this point.

Margaret is the inhabitant of the to-be-ruined cottage: her once happily domesticated family is a victim of failed harvests and enforced unemployment, and her husband, seeing no signs of economic recovery, enlists as a soldier in the war against France and is not seen again. The wholly benign and homely stability of Margaret's former life now turns into an unhealthy waiting/wasting, and such local wandering as takes place is no more than a temporary escape from the whirlpool of enervating depression that centres with morbid persistence on the cottage. Forever looking up the road leading to the cottage in the desperate hope of seeing her returning husband, Margaret's monomania shows the dark side of the romantic appeal of the 'public road' which Wordsworth writes of in *The Prelude*:

> Yet still
> She loved this wretched spot, nor would for worlds

> Have parted hence; and still that length of road
> And this rude bench one torturing hope endeared,
> Fast rooted at her heart, and here, my friend,
> In sickness she remained, and here she died,
> Last human tenant of these ruined walls.   (ll. 522–8)

These are the concluding lines of the poem in the version that Wordsworth completed in 1798, and they refer us back once again to the female beggar in *An Evening Walk*, who also yearned futilely for a returning soldier, and found the path to hope 'choked'.

However, just as *An Evening Walk* ends more positively, with a newly empowered (and personified) Hope illuminating the poet's onward journey, so the melancholy depiction of Margaret's distracted ramblings is not without its counterbalancing elements within the overall structure of *The Ruined Cottage*. Her process of physical and moral decay is narrated by an itinerant Pedlar, a Wordsworthian alter-ego, who represents a positive form of vagrancy: his wanderings are hard and necessitous – walking is literally labour for him – but they are naturalised through repeated association with the seasonal cycle, and they are also the means by which he continually renews his intercourse with an active universe:

> To every natural form, rock, fruit, and flower,
> Even the loose stones that cover the highway,
> He gave a moral life; he saw them feel
> Or linked them to some feeling.   (ll. 80–3)

But even the Pedlar is not an out-and-out unreconstructed nomad, and his relationship with Margaret exposes the inbuilt contradictions of the peripatetic theme: his nostalgic recollection of how he could 'never pass this road' (l. 147) in the past without being given a father's welcome by Margaret neatly expresses the opposing pulls of vagrancy and domesticity in Wordsworth's poetry, and his memories of Margaret would not have the elegiac force that they do (as in the following superb lines) unless there were some regrets attached to his permanently migrant existence:

> She is dead,
> And nettles rot and adders sun themselves

> Where we have sat together while she nursed
> Her infant at her bosom.   (ll. 162–5)

In these lines, and the longer passage from which they are taken, we are offered the same understated disjunction between human interests and a nature obedient to its own imperturbable rhythms as we observed in the preamble to the main story, and which is at the centre of the poem's tragic vision. In aesthetic terms, indeed, the Pedlar is the poet's textual agent and the means by which human suffering and 'natural wisdom' (l. 253) are brought into dialogue: his alternate wanderings and returns to the cottage are a way of obtaining distance and perspective on the tragedy, are a labour of reconciliation; only by repeatedly walking away can the typically Wordsworthian moral – that consolation for human suffering can be found in the peace and beauty of nature – be successfully incubated. This moral, and full philosophical closure, are admittedly only realised in later versions of the poem, but they are implicit throughout the 1798 text, and more than implicit at the end of the first Part, where the 'restless thoughts' inspired by stories such as Margaret's are said to be remediable by 'natural comfort'. What happens when such restless thoughts – the property of what is called, with surely unconscious irony, an '*untoward* mind' (ll. 251–6) – are eliminated entirely, and nothing remains to disturb the 'still season of repose and peace' (l. 246), is best assessed by looking at *Home at Grasmere*, to which I now turn.

Home at Grasmere, another poem belonging uniquely to the posthumously reconstructed Wordsworthian canon, is an autobiographical thanksgiving which records the long walk Wordsworth and his sister took from Sockburn in Durham to take up residence in Grasmere at Christmas 1799. It is a poem written in praise of settled domesticity and the kind of ideal circumscribed freedom which the Vale of Grasmere represents for him. This dominant theme is clearly announced in the opening fifty-odd lines, which find the speaker overlooking the Vale, and remembering having halted in the same place as a boy, and having fantasised about staying forever in this terrestrial paradise: he would enjoy the liberty of a bird or butterfly – 'Creatures that are Lords / Without restraint of all which they behold' – , but one that confines itself to 'flitting' within 'the bounds of this huge Concave' (ll. 32–3, 42).[29] (The dialectal sense of 'flit' as 'remove from one habitation to another' [*OED*], familiar from John Clare's poem 'The Flitting', may be active here: if so, it reinforces one's impression of the drastic curtailment of Wordsworth's vagrant impulse, as his home now shifts only 'from field to rock, from rock to field' [l. 37] within narrow bounds.) When Wordsworth recalls his boyish desire, that 'here / Should be my home, this Valley be my World' (ll. 42–3), he gives us a presentiment of that perfect convergence of the geographical and spiritual centres of one's life that he believes he has now attained.

This message is heavily underlined in the next two hundred lines, which describe him taking imaginative possession of the Vale as a place in which he can take an 'unappropriated bliss' and become the proud 'Lord of this enjoyment' (ll. 85, 87). These phrases recall with some irony the propertyless condition of the Wordsworths: his seigniorial rights are of the heart and mind only. It is a place in which his walks will now begin and end, and which will provide, as he says in the lines (154–70) that attempt to define its singular appeal, a 'termination and a last retreat'. However, just as the lines on his unappropriated bliss are crosscut by rhetorical self-questioning about the 'cost', the 'sacrifice', the semblance of 'conquest', the 'weak indulgence', involved in his acceptance of his life's bounty, so the definition passage is not entirely free of ambivalence: there cannot help but be sombre nuances to words like 'termination' and 'retreat', while a spot 'Made for itself and happy in itself' begs the question of what degree of belonging the human arrivees can reasonably aspire to.

I have said enough to indicate that I am less than entranced by *Home at Grasmere*. The poem has received just attention from modern criticism, and has been highly prized by some commentators: in Anne Wallace's account of peripatetic, for example, it is a key instance of Wordsworth's revision of georgic, whereby he 'represents excursive walking as an agent of home-making and nation-building, and shows the coincidence of its deliberate returns with the recollective action of poetry'.[30] But what sounds energetic and forward-looking in summary is not necessarily so at the level of the reading experience:

> Hail to the visible Presence! Hail to thee,
> Delightful Valley, habitation fair!
> And to whatever else of outward form
> Can give us inward help, can purify
> And elevate and harmonize and soothe,

> And steal away and for a while deceive
> And lap in pleasing rest, and bear us on
> Without desire in full complacency,
> Contemplating perfection absolute
> And entertained as in a placid sleep.   (ll. 388–97)

If this passage encapsulates the superordination of the aesthetics of the beautiful that takes place in *Home at Grasmere*, with its equation of pleasing outward forms with moral tranquillity and its promotion of tender contemplation above desire, it also reminds us of Uvedale Price's warning that beauty requires at least a 'fringe' of that which is other and extrinsic to it – it is this 'which gives it life and spirit, and preserves it from baldness and insipidity'.[31] The placid sleep of Grasmere now seems sugary and inert, a sleep akin to the unending circumrotatory calm undergone by Lucy in 'A Slumber did my Spirit Seal'. The tension between itinerancy and domesticity which makes *The Ruined Cottage* such a fascinating poem is here almost wholly absent: it is difficult to set much store on Wordsworth's assertion that 'the stream / Is flowing and will never cease to flow, / and I shall float upon that stream again' (ll. 383–5), even if a passive surrender to the motion of the stream is accepted as a figure of comparable value to the willed exertions of pedestrian action. In the passage on the slow opening to the newcomer of the 'inward frame' (ll. 693–709), in which Anne Wallace rightly detects the unfolding of pedestrian perspectives as the form of Wordsworth's domesticating vision, his perambulations are identified with the progress of Hope in the manner established as far back as *An Evening Walk*; but his spiritual journey no longer finds its literal ground in the poet's bodily migrations, becoming instead a journey of ever-decreasing circles into the geographical and human heart of his adoptive home.

I therefore (re)turn finally, perhaps with tedious predictability, to *The Prelude* as the acme of Wordsworthian peripatetic in all its thematic resourcefulness and aesthetic complexity. It is, of course, a poem which in its autobiographical and self-referential aspects finds its most natural and compelling metaphors for both the progress of a life and the progress of the poem itself in a pedestrian journey; as John Elder says, 'Wordsworth's understandings of history, of poetry, and finally of the integrity of his own life may all be related to...depictions of himself walking'.[32] To illustrate this one need go no further than the preamble to Book I, with the false optimism of its celebration of the existential and creative freedom of the wanderer and city-escapee who is 'free, enfranchised and at large' (I. 9). Anne Wallace interprets the passage as 'a crucial instance of Wordsworth's ongoing redefinition of "walking" and "wandering"', finding its rhetorical elision of wandering and settlement fulfilling the agenda (walking as cultivation) of the revisionary georgic she credits Wordsworth with introducing.[33] I would argue myself that the exhilarating sense of freedom associated with improvised walking in this passage is conditional upon its being merely *anticipatory* of an eventual resettlement (Kenneth Johnston has interestingly noted the coincidence of Wordsworth's periods of creativity on *The Recluse* with transitional phases in his domestic existence),[34] and that this falls some way short of prefiguring a conflation of walking and writing with productive labour; but perhaps this just reflects my lack of investment in the georgic connection.

While I would tend to respond more, with David Simpson, to that anxious strain in Wordsworth that undercuts his attempts to portray poetry as work,[35] and would argue that this is partly accounted for by the deep interinvolvement of his writing with a peripatetic impulse that is always more than 'excursive' in the range of its ambitions, I would not disagree that walking, whether as material practice or textual figure, is central to the performance of self-cultivation in *The Prelude*. There is much, especially in the early books, to confirm Elder's eloquent observation on how Wordsworth resolves his sense of perplexity over the constant reconstitution of personality through time:

> Wordsworth...finds that to walk back through childhood's landscape is to rediscover the paths 'that first led me to the love / Of rivers, woods, and fields.' Wordsworth's genius is one of understanding apparently hopeless oppositions as stages in a subtle and life-giving *sequence*. The adult's complex emotions derive from 'the simple ways' of the child in nature. For this reason, to walk familiarly over the earth is to be reconciled with oneself in a circuit of rediscovery, wholeness, and belonging.[36]

I think this emphasis on reconciliatory sequence, and on a 'circuit of rediscovery' that is always in process of being remade, is exactly right, and bears closely on some of the textual operations I have examined in Wordsworth's pedestrian poems of the 1790s. I would qualify it only by adding that the fruitfulness of the idea of sequence should not be lost through too eager an invocation of wholeness and unity. To reconcile differences is not to erase them, as Elder himself acknowledges, and he takes the measure of Wordsworth's exploration of subjectivity most accurately when he points up the impossibility of accomplishing any synthesis of the elements of an individual existence:

> Rather, going from one foot to the other, human life takes its passage through a universe of particulars.... Wordsworth's walk... shows him glimpses of a world of resolution [like the Vale of Grasmere, that 'Whole without dependence or defect'?] into which he, while living, can never enter.[37]

Continuing this theme, it would take far more space than I have to follow the twists and turns of the pedestrian metaphor – in the twin dimensions of life-as-a-walk and poem-as-a-walk – through the whole course of *The Prelude*, unravelling the detailed narrative by means of which Wordworth 'converts the wayfaring Christian of the Augustinian spiritual journey into the self-formative traveler of the Romantic educational journey' (in the words of M. H. Abrams),[38] to its culmination in the ascent of Snowdon in Book XIII. I intend instead to focus my discussion on Book VI, the section that deals with the same walking tour of France, Switzerland and Italy as *Descriptive Sketches*, and which includes the well-known passage on the crossing of the Alps. Book VI recapitulates a number of the themes I have explored in my discussion. It is a part of the poem which progresses with a pedestrian rhythm of varying pace and variable continuity, and which subtly interweaves sensation, observation, reflection and memory. It provides a conspectus of the personal and social significances of Romantic pedestrianism as well as demonstrating its most productive and innovative aesthetic uses.

As the part of the poem which witnesses the most dramatic intersection of the Imagination (in the famous apostrophe of lines 525–48) with pedestrian action (in the form of the unrestrained mobility of the European walking tour), it is not surprising that Book VI should closely associate walking with that period in Wordsworth's life at Cambridge when his poetic ambitions first took definite shape and 'the dread awe / Of mighty names was soften'd down' (VI. 72–3). Even the most incidental phrases – 'The rambling studies of a truant Youth' (VI. 111), 'I had stepp'd / In these inquiries but a little way' (VI. 137–8) – show the extent to which Wordsworth's understanding and representation of his past are mediated by the peripatetic metaphor. He also places the dawning of his poetic self-confidence explicitly in the context of his daily evening walks. He describes one particular picturesque, ivy-clad ash tree, before which he habitually 'stood / Foot-bound' (VI. 100–1) in the course of these walks. The picturesque gaze at first seems categorically to arrest pedestrian progress, but the continuation effects the familiar modulation from physical into mental walking:

> The hemisphere
> Of magic fiction, verse of mine perhaps
> May never *tread*; but scarcely Spenser's self
> Could have more tranquil visions in his youth,
> More bright appearances could scarcely see
> Of human Forms and superhuman Powers,
> Than I beheld, standing on winter nights
> Alone, beneath this fairy work of earth. (VI. 102–9; italics added)

The reference to Spenser is one of the miscellaneous acts of self-identification with his revered quadrumvirate of precursors (Chaucer, Spenser, Shakespeare and Milton) that Wordsworth makes in his poetry and letters, and although he hesitates to embrace 'magic fiction' as his destiny it is clear that he intends to 'tread' some other realm of poetic discourse, and that the expression of this ambition via a pedestrian image carries no negative connotations.

A distinct minor lexical vein associated with walking in Book VI is furnished by the semantically-related verbs, 'scatter' and 'spread'. With their powerful agricultural connotations, they are verbs that give shadowy presence to the *Georgics*, taken by Wallace to be the leading influence on Wordsworthian peripatetic. Without wishing to go nearly so far, these images do offer an interesting sidelight on the results of that persistent intertwining of 'self-wandering' and 'poem-wandering' (to borrow some apt phrases from Kenneth Johnston) that is so conspicuous in Book VI. The first incidence is in the section describing his wanderings in the Yorkshire Dales and Lake District with Dorothy and Mary Hutchinson: over all the ground they covered together, thereby endeared by shared experience, 'was scatter'd love, / A spirit of pleasure and youth's golden gleam' (VI. 244–5). This fertilising of the circle of close relationships then breeds effects in the textual present, as the agricultural metaphor is extended to the absent addressee, Coleridge:

> O Friend! we had not seen thee at that time;
> And yet a power is on me and a strong
> Confusion, and I seem to plant Thee there.   (VI. 246–8)

The second instance comes in the early stages of the account of the walking tour, when Wordsworth is describing the mood of national joy in which he and Jones participated as they travelled through France a year after the fall of the Bastille: on both the public roads and country paths they 'found benevolence and blessedness / Spread like a fragrance everywhere' (VI. 368–9). The blessedness is spread most obviously by the emancipated French, although as Wordsworth goes on to recall how the two tourists were welcomed and entertained, by a group of *fédérés* returning from Paris, 'As their forerunners in a glorious course' (as free-born Englishmen, that is), he all but claims credit for the auspicious time for themselves. The third example comes in the final reflections on the summer in France, and shows a more straightforward self-representation as the 'cultivator' of a pleasure higher than that of revolutionary triumph:

> the ever-living Universe,
> And independent spirit of pure youth
> Were with me at that season, and delight
> Was in all places spread around my steps
> As constant as the grass upon the fields.   (VI. 701–5)

What the sequence of agricultural images demonstrates compactly is how human relationships, and even the great tide of collective history, are subordinated in terms of their creative, life-enhancing potential to the mind of Wordsworth himself. That mind is at once cultivator and cultivar, a source of fertile imaginings and the object of Nature's husbandry, as one would expect to see it represented in a poem on the *growth* of a poet's mind.

If Wordsworth depicts himself literally as a serious pedestrian, and more figuratively as a walker-labourer, a cultivator, the poem too borrows its rhythms from those of pedestrian travel. 'I, too, have been a Wanderer' (VI. 261), Wordsworth states after alluding to Coleridge's current convalescent stay in Malta; finishing his account of the walking tour, he cautions himself, 'I must break off, and quit at once, / Though loth, the record of these wanderings, / A theme which may seduce me else beyond / All reasonable bounds' (VI. 658–61). The rhetorical overlap between the wanderings of the biographical self and the divagations of the text is especially acute here, but is a meaningful presence throughout the Book. Book VI, in common with *The Prelude* as a whole but in a more palpably overdetermined way, *is* a walk – along a path that is sometimes direct, but more often 'circuitous' (VI. 680) – , and it needs to be emphasised that this is more than a loose analogy. The features of this Book that I have drawn attention to so far are the distinctive features of Wordsworthian peripatetic, and they flow unmistakably from the bodily and mental experience of pedestrian motion. The lines referred to above in which Wordsworth confusedly 'plants' Coleridge in scenes where he was not present are followed by a more self-conscious fantasy in which the two were contemporaries at Cambridge – which Wordsworth assumes would have been to Coleridge's benefit, though he admits that the latter 'hast *trod* / A march of glory, which doth put to shame / These vain regrets' (VI. 327–9; italics added). The kind of rememorising indulged in here helps to make sense of a passage that has puzzled some critics – the added lines in the 1850 *Prelude* in which Wordsworth describes witnessing the expulsion of the monks from the convent of the Grande Chartreuse by the revolutionary army, though this did not actually happen until two years after his walking tour. The anachronism takes its place readily in a Book that deals with several such re-imaginings: Wordsworth 'sees' the desecration in the same way that he 'sees' Coleridge with Dorothy and Mary in the Lake District.

I am suggesting that the 'free association' or recombination evident here, which was observable also in the *Salisbury Plain* poems, is one of the properties of peripatetic form. The leisurely pace, the detours, pauses and returns, the slow unfolding of perspectives, the surprises and contingent pleasures, the mixture of weary inattention and heightened awareness, the disordering of one's sense of time, the space for reflection on what one has seen and experienced and the beginning of the filtering operations of memory – all of which will be recognised as part of the texture of walking (especially the more improvised, 'outward-bound' kind of walking) by anyone who has participated in the activity: all these have their intimate correlates in the unhurried and circuitous progress, the 'variegated journey' (VI. 427), of Wordsworth's verse, its elaborations and digressions, its variable intensity, its sequencing and recombining, its loosening of spatial and temporal fixities, its unobtrusive rhetorical patterning. These properties have to be understood in relation to each other: only an inconsolable dualism of mind and body would insist on seeing one as a faint and inessential analogy of the other.

Something more must be said of the narrative of the walking tour in Book VI, and in particular of its centrepiece, the crossing of the Alps, that has dominated a generation of criticism of *The Prelude*. The suppleness of the verse in capturing the 'variegated' experience of the 'two brother Pilgrims' (VI. 478) is remarkable. It is particularly successful in representing the perceptual and mental states that accompany the movement *between* places – not the exhilaration of departure, or the pleasures of arrival, but what Eric Leed calls the 'experience of passage', that is almost illegible in an entire tradition of travel writing:

Where Elms, for many and many a league, in files,
With their thin umbrage, on the stately roads
Of that great Kingdom, rustled o'er our heads,
For ever near us as we paced along,
'Twas sweet at such a time, with such delights
On every side, in prime of youthful strength,
To feed a Poet's tender melancholy
And fond conceit of sadness, to the noise
And gentle undulations which they made.
. . . .
A march it was of military speed,
And earth did change her images and forms
Before us, fast as clouds are chang'd in Heaven.
Day after day, up early and down late,
From vale to vale, from hill to hill we went
From Province on to Province did we pass,
Keen Hunters in a chace of fourteen weeks.... (VI. 371–9, 427–34)

The alternately tranquillising and stimulating effect of passage; the sense of a rush of impressions, and the telescoping of time, which may not affect the pedestrian in the present but may well do *in retrospect*; the way in which, as Leed puts it, 'The passenger becomes more conscious of self as a "viewer" or "observer" of a world flowing past':[39] these qualities are all memorably inscribed. Even the apparent incongruity of the poet's aesthetic posturing (his fashionably 'tender melancholy') with the otherwise 'perpetual hurry of delight' strikes one as authentic, in that it suggests the way consistency of mood and personality have been surrendered to the pleasure in motion and change.

Equally, Wordsworth's verse is capable of deviating from the habitually loose syntactic patterns of peripatetic to register moments of particular intensity and imaginative power. These can occur as *apparent* interruptions of passage, as in the Gondo Gorge section, where the 'slow step' of the travellers seems further retarded by the seventeen-line sentence beginning 'The immeasurable height of woods decaying', in which what are presented as objects in grammatical sequence are more properly the elements of a simultaneous impression. They can also occur at times when motion is actually arrested, as in the following passage (which in some ways, as Kenneth Johnston observes, seems to parody the Gondo section), which describes the night when Wordsworth and Jones mistakenly get up at 1.45 a.m., get lost in a wood and have to sit it out till dawn:

> the cry of unknown birds,
> The mountains, more by darkness visible
> And their own size, than any outward light,
> The breathless wilderness of clouds, the clock
> That told with unintelligible voice
> The widely-parted hours, the noise of streams
> And sometimes rustling motions nigh at hand
> Which did not leave us free from personal fear,
> And lastly the withdrawing Moon, that set
> Before us, while she still was high in heaven,
> These were our food.... (VI. 644–54)

Here, the long periodic sentence, the sustained parataxis, and the *synathrismos* or heaping-up of noun-phrases denoting sounds and sights, help to convey great sensory pressure bordering on hallucinatory self-deception (the 'widely-parted hours', the moon that appears to set by dropping behind the mountains), and generate considerable rhetorical tension that is released rather anticlimactically by the banal main clause, 'These were our food'. As Johnston notes, this experience, like the Italian clocks, seems largely unintelligible, offering no compensatory moment of visionary insight,[40] and its inclusion, I would add, is testimony partly to the inclusiveness of peripatetic, its dedication to sequence rather than synthesis. It is perhaps significant that it is here that Wordsworth breaks off his narrative of the tour, realising that it will 'seduce' him 'beyond / All reasonable bounds' (VI. 660–1). I am not denying absolutely the existence of such patterns as Johnston, Mark Reed and others perceive in this whole section of Book VI:[41] the pattern of expectation-confusion-clarification ascribed by Johnston and Reed to the French, Swiss and Italian stages would probably occur to most readers, though the symmetries are scarcely tidy. But to present this as an 'overall movement from thesis to antithesis to synthesis'[42] is, I think, to give a misleading impression of conceptual order. Certainly the sequential movement of the narrative provides opportunities for repetition, comparison and reinterpretation, just as a walking tour will constantly invite comparisons between scenes and experiences along the way. But the passages Johnston sees as the final phases of the poem's dialectical structures are never more than resting-stages on the text's circuitous path. John Elder's comments on *The Prelude*'s 'circling dialectic that yields no synthesis', and on 'reconciliation, rather than resolution'[43] as the keynote of the poem, are worth recalling here: indeed, if the account of the walking tour has a thematic pattern it is one wherein the mind is frustrated in the trials of its 'youthful strength' (VI. 376) and 'reconcil'd...to realities' (VI. 461), but for this to become a pattern at all the reconciliation must be short-lived.

This entire process is fully legible in the Simplon Pass episode, which, overfamiliar as it is, I shall refrain from quoting extensively. The disappointing discovery that, after mistaking their path, he and his companion have unwittingly crossed the Alps, leads Wordsworth to interrupt his very matter-of-fact narrative and address the Imagination as a faculty that glories in infinite desire. This visionary moment is couched in a pedestrian metaphor – an example of what Geoffrey Hartman sees as an archetypal image in Wordsworth's poetry, that of the 'halted traveller'' (legible in the 1805 *Prelude*, but more explicit in the 1850), the significance of which he defines as 'consciousness of self raised to apocalyptic pitch'.[44] However, both the depression caused by the peasant's news that they have crossed the Alps (in the narrative past), and the paralysis of Imagination's self-recognition (in the textual present), are immediately dislodged by renewed physical or poetic progress. With existence understood as process, as 'something evermore about to be', any moment of vision can only be self-cancelling – a temporary halt on the trail of a life or a text. For Hartman, in the Gondo Gorge section that follows Wordsworth reaffirms his filial relation to nature and the mimetic function of his poetry: 'He moves haltingly but he moves', sure that 'the way is the song', though the song had threatened to become the way.[45] My reading of *The Prelude* as peripatetic suggests rather that Wordsworth manages to have his cake and eat it: the way is the song *and* the song is the way, and I see no unbridgeable gulf between the two.

Nevertheless, one can agree with the broad thrust of Hartman's reading, that in the descent into the 'gloomy Pass', where the travellers enter a scene of 'tumult and peace' and 'darkness' and 'light', and natural features are construed as 'The types and symbols of Eternity', Wordsworth recoils from the apocalyptic potential of his imagination and misrecognises it as an apocalypse of Nature. To pick up the terms of my earlier discussion, one might redirect this analysis by showing how motion and stasis are projected into nature as workings of the one heavenly mind, rather than the one *human* mind in its oscillation between the restlessness of desire and the reassurance of stable identity. However, if the Gondo Gorge is read as the resolution of the halted traveller's predicament, as the reconciliation of Imagination and Nature, then this reconciliation itself does not last long, as the prosaic cares of the pedestrian traveller supervene in a curious tailpiece to the whole episode:

> That night our lodging was an Alpine House,
> An Inn, or Hospital, as they are nam'd,
> Standing in that same valley by itself,
> And close upon the confluence of two Streams;
> A dreary Mansion, large beyond all need,
> With high and spacious rooms, deafen'd and stunn'd
> By noise of waters, making innocent Sleep
> Lie melancholy among weary bones.   (VI. 573–80)

It is a recapitulation in a more homely register. The detail of the 'confluence of two Streams' gestures back irresistibly to the fellowship of brook and road in the gloomy pass and the 'confluence' of human and natural it stands for, while the 'raving stream' and 'stationary blasts of water-falls' in the 'immeasurable height' of the Gorge now echo eerily in the 'high and spacious rooms' of the Gondo Spittal. Though Nature may at times seem fitted to the mind, here neither the natural nor built environment make much allowance for the body. 'Apocalypse is not habitable', as Hartman wryly observes,[46] and the night in the dreary mansion serves to underline that fact: it is with some relief that the two pedestrians renew their journey the following day.

It seems wise to 'break off...the record of these wanderings' here, with a few summary reflections. The truth of Wordsworth's tale is in the 'wandering' of its argument, after all, and there is a sense in which a critical account of that argument must respect its mode of passage, and not provide the kind of false syntheses that the poet himself avoids. I have drawn attention throughout this chapter to the two poles of a complex sensibility: we could call these the itinerant and the settled, the vagrant and the domestic, the free and the bounded. Although the initial terms of these oppositions most closely and obviously respect the dynamic of a peripatetic art, Wordsworth is at his best as a pedestrian poet, to return to Seamus Heaney's elliptical judgement, when he promotes neither term to the exclusion of the other but honours instead the dialectical tension between the two. As I have tried to indicate in my commentaries, each of the two poles is also susceptible of both a positive and a negative inflection. The itinerant vision receives its most affirmative embodiment in such figures as the Pedlar in *The Ruined Cottage* and the autobiographical persona in *The Prelude* – that is, the homeless, jobless, physically and spiritually uprooted Wordsworth of the 1790s. But it requires only a loss of self-confidence for itinerancy to reveal its darker side, and when harnessed to an explicit social or political agenda it mutates into the assorted vagrants, beggars and outcasts panting up the endless Alp of life. The opposing vision of a happily bounded existence is repeatedly sketched in in the early poems – in the pastoral idyll of Alpine shepherds in *Descriptive Sketches*, for instance – and gets its most sustained and enthusiastic endorsement in *Home at Grasmere*. But this vision also suffers an inversion: even if *Home at Grasmere* is read as a wholly successful act of praise, there is the powerful image of Margaret's slow stagnation in the ruined cottage to countermand the leaning towards overvaluation of particular places, or, more generally, towards a placid life in a small, protective community.

I have made it clear that, in my opinion, for all the considerable human, social and aesthetic interest of the earlier poems, it is in *The Prelude* that this complex sensibility finds the equally complex and sustained expression it requires, in the form of the autobiography of a great pedestrian that becomes itself a feat of long-distance textual pedestrianism. In a stunning short poem, 'Walking', by the seventeenth-century devotional poet Thomas Traherne, we are told:

> To walk is by a thought to go;
> To move in spirit to and fro....

Wordsworth's poetry, and pre-eminently *The Prelude*, offer a rich elaboration of Traherne's early insight.

5
'Indolence Capable of Energies': Coleridge the Walker

In his own time, Coleridge's manner of walking was a noteworthy feature of the physical impression he made on others, and there was a temptation to draw intellectual or moral inferences from it. Coleridge himself commented, in a letter to Thomas Poole, on his 'awkward' gait, and of its indicating 'indolence capable of energies'. William Hazlitt remarked upon Coleridge's habit of crossing from one side of the footpath to the other in front of him, and connected this both with his fluid, digressive conversational style and with his 'instability of purpose' or 'involuntary change of principle'. In Chapter 4 I recalled Hazlitt's memorable comparison of Wordsworth's and Coleridge's habits of peripatetic composition – Coleridge preferring to compose in 'walking over uneven ground, or breaking through the straggling branches of a copse-wood'. A further familiar anecdote is that of Coleridge's arrival at Racedown in Dorset (then the home of Wordsworth and his sister) in 1797: Coleridge, Wordsworth later remembered, 'did not keep to the high road, but leapt over a gate and bounded down the pathless field, by which he cut off an angle'. This recalls a passage in a letter Coleridge wrote to Robert Southey during his pedestrian tour of Wales in 1794: he celebrates the coining of a word by denouncing the 'circuitous, dusty, beaten high-Road of diction' in favour of cutting across 'the soft, green pathless field of Novelty'.[1] If body, as Coleridge suggests in a late essay, is 'but a striving to become mind', it seems legitimate to take seriously these gestures at linking bodily action, intellectual and moral attitudes or tendencies, and aesthetic form, and this is one aim of the present chapter.

Coleridge's early life offers a rich and colourful portrait of one who adopted pedestrianism initially in the spirit of someone subscribing to a fashion, albeit one that satisfied him on doctrinal grounds, but for whom walking quickly became a vital adjunct of his creative life. The pedestrian tour of Wales in July/August 1794 was followed immediately by a walking tour of Somerset with Southey, during which they continued their Pantisocratic planning, and a year later they were falling out over the same subject during a walk up the Wye valley with Joseph Cottle.[2] Coleridge walked over 60 miles in three days from Bristol to Racedown in Dorset in 1797 to visit Wordsworth, and, having persuaded him to move to Somerset, spent much of the next year exploring the Quantocks on foot with him and Dorothy. The implication of some of Coleridge's most memorable poetry in these pedestrian expeditions is a matter of record: 'Kubla Khan' was composed in the course of a long unaccompanied coastal walk to Lynton and back; 'The Ancient Mariner' was conceived during a walk of several days to Dulverton in the company of Wordsworth and Dorothy. Coleridge's letters, Dorothy's journals, Hazlitt's recollections, and other familiar sources, construct a picture of incessant, and physically very demanding, excursionary walking over this period, during which the major Conversation Poems were written and 'Christabel' was begun. This activity continued during his stay in Germany, despite the intensiveness of his scholarly researches, and included a week-long pedestrian tour of the Harz mountains in May 1799, described in meticulous detail in his Notebooks and in letters to his wife. In the three years after returning to England, he undertook a walking tour of Devon with Southey, a three-week tour of the Lakes with Wordsworth and his brother John, and – the highpoint of his own residence in the Lake District – a nine-day, 100-mile fellwalk around the region in August 1802, taking in the summit of Scafell.[3] A year later Coleridge joined William and Dorothy for a tour of Scotland in a one-horsed 'jaunting car', but took an early opportunity to leave the 'Hypochondriacal' and '[self]-centered' Wordsworth and continue his journey on foot, which amounted to an impressive '263 miles in eight days'.[4] Soon after Coleridge left the Lake District, and in April 1804 he sailed for Malta, an event which effectively marks the end of his remarkable career as a pedestrian.

My highly telescoped summary of that career will at best have demonstrated that walking was very much part of the texture of Coleridge's life over a significant period; the more interesting and challenging task is to determine how these biographical facts, and their relation to his writing in various forms and genres, should be interpreted. A crude summary might be that the decade 1794–1804 takes Coleridge from a radical politics of walking to a radical poetics of walking.

I have already discussed, in Chapter 2, the anticonformist dimension to the pedestrian tour in the 1780s and 1790s, and talked briefly of Coleridge's and Hucks's Welsh tour in 1794 as an exemplary case of 'radical walking'. The most celebrated incident of the tour is one which nicely expresses the politicised climate in which it was undertaken. By his own account, Coleridge precipitated a fracas at an inn in Bala by proposing a Republican toast, though he gives the subject of the toast as Washington in one letter, Priestley in another. Hucks, in his published tour, reports the toast and the resulting 'violent and dreadful battle of tongues' ('hailstorms of Vociferation', in Coleridge's phrase), but ascribes it to an anonymous third party. It seems reasonable to suppose, as Hucks's modern editors do, that this was done to protect either Coleridge or himself from accusations of sedition.[5] Hucks generally comes across as a typical composite of the philosophical traveller seeking to 'make some observations upon the human character' in a region seemingly conducive to Rousseauesque deliberations on the natural virtues of 'unpolished people', and an aesthetic questor of the 'hidden beauties of nature unmechanized by the ingenuity of man', which he approaches with a conventionally pictorial eye.[6] Nevertheless, he manages to convey his prudently restrained radicalism both directly, as in his condemnation of the Treason Trials in Scotland and the suspension of Habeas Corpus, and indirectly, in politicising the appreciation of picturesque ruins, as at Carnarvon:

> Every castle that now remains is a monument of shame to our ancestors, and of the ignoble bondage under which they bent: and hence in part arises that satisfaction...in contemplating their ruins....[7]

Coleridge, having devised with Southey the plan for a communitarian utopia on the banks of the Susquehanna, tells his fellow Pantisocrat that he proselytised to good effect in Llanvillin, causing 'two great huge Fellows, of Butcher like appearance' to dance round the room in excitement'.[8] The letters he wrote on the tour convey vividly the energy and mobility of a young radical mind, discovering in movement through space a release from the defining social contexts of his life to date and a realm of physical difference onto which his quest for self-realisation could be mapped. Travel, indeed, would continue to benefit Coleridge in such a way, long after his retreat from political activism, but the buoyantly Republican mood of the 1794 tour seems to have enhanced the effect. For him, the vitalising 'experience of passage' is translated into the linguistic exuberance that he displays and comments on in his first letter to Southey: 'as I journey onward, I ever and anon pluck the wild Flowers of Poesy – "inhale their odours awhile" – then throw them away and think no more of them'; and this indulgence in disposable similes is matched by a degree of mental and emotional weightlessness, whereby he can pass rapidly from reproving Southey's despondency, to 'melancholizing' on his own account when he encounters his first love, Mary Evans, to declaring breezily that 'I am 16 miles distant, and am not half so miserable'.[9]

These positive effects of travel are partly offset by those factors that register most unmistakeably the pedestrian mode of the tour, namely the bodily fatigue and the unique sensitivity to sensory stimuli, unpleasurable as well as welcome. Coleridge writes of the intolerable 'heat and whiteness of the Roads', of '*stone*-fences dreary to the Eye and scorching to the touch', and vividly displaces the intermittent strains of pictorial detachment in his landscape descriptions by referring to the most intimate oral contact with rugged mountain terrain: 'I applied my mouth ever and anon to the side of the Rocks and sucked in draughts of Water cold as Ice, and clear as infant Diamonds in their embryo Dew!'[10] The intense heat and dazzling whiteness of the roads and surrounding scenery carry intimations of the sublime – Burke had said that 'such a light as that of the sun, immediately exerted on the eye, as it overpowers the sense, is a very great idea'[11] – but Coleridge typically does not shrink in Burkean subjection on that threshold: here, instead, he disarms the sublime object by ingesting it, finding that the 'stupendous' rocks about him have their own gentler, life-sustaining aspect. The same limited (in range and complexity) yet concentrated perceptions as those just noted furnish the texture of the best of the few poems written on the Welsh tour, 'Perspiration, a Travelling Eclogue'. Here they also provide context for Coleridge's discrimination between carriage-travelling aristocrats and the implied pedestrian speaker:

> The Dust flies smothering, as on clatt'ring Wheels
> Loath'd Aristocracy careers along.

> The distant Track quick vibrates to the Eye,
> And white and dazzling undulates with heat.
> Where scorching to th'unwary Traveller's touch
> The stone-fence flings it's narrow Slip of Shade,
> Or where the worn sides of the chalky Road
> Yield their scant excavations (sultry Grots!)
> Emblem of languid Patience, we behold
> The fleecy Files faint-ruminating lie. – [12]

The lazily ornamental diction of 'fleecy Files' is a mercifully brief deterioration (a return to the 'beaten high-Road of Diction') in a poem that begins so impressively, and which confirms Coleridge as the 'sturdy Republican' he addresses Southey as. His companion, evidently, was a radical walker of more impure faith, as Coleridge suggests in reporting Hucks's dismay at seeing a 'little Girl with a half-famished sickly Baby' put her head in at the window of an inn – dismay not at her plight but at having the 'clamorous Voice of Woe' intrude on his ear at dinnertime.[13]

In turning to Coleridge's later work in prose and poetry of the 1790s, the politics of walking effectively disappears as a theme, but a rich and diverse literature presents itself as material for a poetics of walking very different to, and in certain ways more innovative than, that which I ascribed to Wordsworthian peripatetic in Chapter 4. To begin with, Coleridge's notebooks and letters narrating and annotating his extensive walking in Germany, the north of England, and Scotland deserve appraisal as travel writing and for the kinds of attention they bring to the natural world. This debate has been initiated by Patricia Ball, Raimondo Modiano, Bill Ruddick and Harold Baker,[14] among others, and my own concern is to co-ordinate it more closely with a study of the physical and psychological conditions of pedestrian travel.

The critics I have cited differ in their total evaluation of Coleridge's landscape descriptions in the *Notebooks*. Baker, who provides the most searching formal analysis, writes of their 'exceptional character', points to their startling non-convergence with the 'normative organic poetics' more readily associated with Coleridge, and argues somewhat puzzlingly for integrating them with a consideration of the 'roads not taken' by Romanticism.[15] Modiano complains – not without reason – that 'they lack the proper organization, both as regards the sampling of relevant information and authorial directions in the form of rhetorical pauses or emphases, by means of which a reader can comfortably follow the observations of the experienced traveller', and criticises as 'breathlessly cinematic' the same technique Ruddick admires, whereby Coleridge offers 'a moving panorama in which the spectator's head is free to move this way and that, and his body to turn about as he advances into a landscape'.[16] All commentators are agreed, however, that in this latter respect, as well as others, Coleridge decisively renounces the conventions of the picturesque gaze. Although I argued in Chapter 2 for a greater congruence between the picturesque and pedestrianism as cultural practices than might at first appear, especially with regard to the resonance of picturesque playfulness with the free mobility of the pedestrian, I accept that Coleridge achieves effects in his prose dealings with landscape quite unlike anything to be found in Gilpin or his followers. Indeed, it may be that Coleridge realises some of the abridged potential of picturesque theory, which picturesque writing, owing to its interdependent relationship with one of the earliest varieties of mass travel, necessarily suppresses, replacing the creative underdetermination of picturesque experience with a repertoire of aesthetic standards and models that can be easily marketed and communicated to, and re-enacted by, the travelling public.

Ruddick's comments (quoted above) indicate that Coleridge's freedom from picturesque conformity stems from a physical immersion in the landscape that can arise only from walking through it. One neat illustration of this is Coleridge's habit of referring to viewpoints as 'Resting Places', in place of the customary picturesque term, 'station'.[17] The fondness for panoramic description that Ruddick usefully highlights is arguably also a pedestrian predilection, not only in the sense that it requires complete freedom of bodily movement and the possible circumambulation of a felltop, but also in that it represents a maximisation of benefit from the physical negotiation of difficult terrain, instead of the easeful contemplation of a prepackaged view by one who has been transported to a location in a carriage or on the back of a pony. Coleridge does not use the specialised thesaurus of picturesque word-painting: instead, as Modiano suggests, he carries 'topographic richness', by which we might understand the intricacy and variety of Price's picturesque theory, to the point of overload,[18] baffling as much as it excites the reader's desire for natural or aesthetic unity. Where he explicitly invokes the picturesque – as he does, for instance, at the end of his account of his entrance into Wasdale during the August 1802 tour, following his description of front and back views with reference to '4 one horse carts in a file on the top of the Fell to my Left'[19] – it seems ironically self-cancelling. It certainly looks, in fact, as though Coleridge's approach to landscape description in his *Notebooks* can be best defined *against* the canons of the fashionable aesthetic of the time. (In passing, though, I have seen no plausible explanation offered of why many of his verbal descriptions and sketch-maps transpose left and right – are, in effect, mirror-images.[20] Might it just be possible that Coleridge was using that indispensable accessory of the picturesque tourist, the Claude Glass?)

The pedestrian's experience of passage is recorded in other ways in these writings, though the persistence of the body as the limit or measure of sensory and imaginative reality is a unifying factor. Coleridge is nothing if not explicit about the co-action of mind and body, as in the following description of rugged terrain witnessed on the Harz tour of 1799:

> Hills ever by our sides, in all conceivable variety of forms and garniture – It were idle in me to attempt by words to give their projections & their retirings & how they were now in Cones, now in roundnesses, now in tonguelike Lengths, now pyramidal, now a huge Bow, and all at every step varying the forms of their outlines; or how they now stood abreast, now ran aslant, now rose up behind each other or now, as at Harzburg, presented almost a Sea of huge motionless waves too multiform for Painting, too multiform even for the Imagination to remember them yea, my very sight seemed *incapacitated* by the novelty and Complexity of the Scene. Ye red lights from the Rain Clouds! Ye gave the whole the last magic Touch! I had now walked five & thirty miles over roughest Roads & and had been sinking with fatigue but so strong was the stimulus of this scene, that my frame seemed to have drank in a new vitality; for I now walked on to Goslar almost as if I had risen from healthy sleep on a fine spring morning: so light and lively were my faculties.[21]

There is no separation here of mind and body, no suggestion of the spirit profiting at the expense of its enfeebled fleshly vehicle; rather, the body responds to the altering health of the mind, visual and aesthetic stimuli apparently bringing revitalisation of the walker's battered 'frame'. That the mutual determination of mind and body is fully reciprocal is indicated by the foregoing lines, which make plain that the overwhelming 'novelty and Complexity' of the landscape proceed exactly from the conditions of pedestrian mobility: 'all *at every step* varying the forms of their outlines...'. It is interesting that, once again, what seems the prelude to sublime crisis – a paralysing incommensurability of mind and nature – is effortlessly smoothed away by invoking, in the harmonising power of light and colour, the white 'magic' of the beautiful. It is also remarkable how Coleridge, in ascribing such energetic variability to the landscape, employs a mixture of geometric figures ('Cones', 'roundnesses', 'pyramidal') and humanising metaphors ('tonguelike Lengths', 'stood abreast', 'ran aslant'). Reading the landscape in terms of the body, so common in the nature writing of his letters and notebooks, seems less a matter of rhetorical convention than of the very insistence in consciousness of the sensations of a body in motion: the ego of Coleridge's travel writing seems very much, in Freudian phrase, a bodily ego, and seeks bodily identifications with the land it traverses.[22] One final point to note is the breathless fluidity of Coleridge's images, the 'multiformity' of the hills as they are perceived in motion from one moment to the next rendered in a freely extendable metonymic series. It is a rhetoric constructed from what Modiano defines as 'an elliptical language made of a string of paratactic sentences and dominated by nouns, adjectives or prepositions' ('now in Cones, now in roundnesses, now in tonguelike Lengths...*etc.*'), and it has the disorientating effect, firstly, of making it impossible to visualise any particular prospect, but achieves the rarer and more interesting result of representing the transformations of passage. This reflects Patricia Ball's shrewd distinction between Coleridge's and Dorothy Wordsworth's nature descriptions, namely that Dorothy is interested in the changes she observes in nature from one season to the next, whereas Coleridge is more interested in changes observed from one *moment* to the next.[23] In the passage above, the multiplication of perspectives and accompanying figurations places extra strain on the eventual discovery of an aesthetic 'whole', suggesting perhaps that Coleridge is not concerned with a conventional pictorial unity, but rather with a unity that is the synthesis of temporally sequenced perceptions, and which can only be imaginatively apprehended.

This latter point could be elaborated in complementary ways. The delight in a various, energetic, changing landscape, in the crowding of visual and mental space with heterogeneous objects (however inexorably this leads to the antithetical wish for aesthetic closure), perhaps has its bodily or locomotive analogue in the observation of Hazlitt's which I noted at the beginning of this chapter: that is, Coleridge's preference for 'walking over uneven ground', for a landscape that offered resistance to the body-in-motion. This can be linked to the more metaphysical concerns explored in Michael Cooke's stimulating essay on the idea of space in Coleridge's work – particularly his poetry, where, Cooke argues, recurrent images of the 'unrelieved plain' and 'chaotic ocean' express proto-existentialist anxieties about the limbo of existence.[24] Cooke usefully homes in on some difficult ruminations on space, time and motion from Coleridge's notebooks of December 1803:

> I believe, that what we call *motion* is our consciousness of motion, arising from the interruption of motion = the acting of the Soul resisted./.Free unresisted action (the going forth of the Soul) Life without Consciousness, properly infinite, i.e. unlimited – for whatever resists, limits, & vice versa/This is (psychologically speaking) SPACE. The sense of resistance or limitation TIME – & MOTION is a Synthesis of the Two. The closest approach of Time to Space forms co-existent Multitude.[25]

Coleridge makes clear in the lines preceding those quoted that he is dealing with what he terms 'psychological', rather than physical, motion: psychological space is associated with the potentially infinite expansion of the soul, appropriating everything effortlessly to its own nature, whereas time is identified with the experience of interference from the material world, the (internalised) encounter with the otherness of objects that circumscribe and help define the self. Unobstructed space, in this sense, seems to be a theoretical desideratum, at best, for Coleridge, but a practical impossibility, and something to which he was emotionally disinclined: however passionately he sought incorporation into an infinite oneness, his transcendental yearnings mean little except in opposition to the prosaic experience of the temporal, finite self – to which they inevitably return. Coleridge wrote that he would 'make a pilgrimage to the burning sands of Arabia...to find the Man who could explain to me there can be *oneness*, there being infinite perceptions'.[26] He found this infinity of perceptions exaggerated by the sensuous experience of pedestrian travel, and seems to have found it exciting and therapeutic, a temporary escape from the 'unrelieved plain' of existential isolation. (It is ironic that Coleridge expresses his yearning for oneness via an image of travel – a foot-journey – that could only defer its realisation.) Just as he preferred his physical path to be obstructed, Coleridge desired the passage of his eye across the landscape to be continually arrested, and the sublime questing of his soul to be materially impeded.

In the context of fiction, description, where it does not function as a reality-effect, is often taken to stand in some figurative relation to plot. But in Coleridge's travel writings, as Harold Baker shrewdly points out, description *is* plot, in the sense of a dual anti-narrative of 'the describer's motion over the surface of the earth' and the movement 'of perception itself in its continual constitution of the world'.[27] This is narrative with a single agent, and with no immanent finality or goal. But Coleridge seems to have welcomed these opportunities to remake the self both through the constant renewal of perception, and through the discouse articulating those perceptions. Baker writes that he inhabits 'a continual present-tense of pure phenomenality',[28] although this is tantamount to abandoning the postulate of (self-)presence altogether, as Coleridge himself jokingly concedes in a letter written on the Scottish tour of 1803:

> For the last 5 months of my Life I seem to have annihilated the present Tense with regard to place – you can never say, where *is* he? – but only – where *was he*? where *will* he be?[29]

Annihilating the present tense is also to remove the ground of stable selfhood, suspending the subject between difference and deferral, memory and anticipation. However much Coleridge may have yearned for secure anchorage in his personal life, there was a parallel and contrary need to remove himself from all localising, defining associations and commitments, to re-experience what he calls (in reflecting on his passion for solitary travel) 'a sort of *bottom-wind*, that blows to no point of the compass, & comes from I know not whence, but agitates the whole of me'.[30] The quintessentially Romantic character of this image of an interior breeze, at once origin-less and originative, emphasises how the experience of passage was for Coleridge intimately allied to creativity in the intellectual or artistic sphere.

Coleridge glosses his reference to the 'bottom-wind', with its connotations of an unconscious drivenness about his walker's appetite for natural scenery, by suggesting that his soul 'must have pre-existed in the body of a Chamois-chaser'. Whimsical this may be, but the supposition of a material cause for habits of feeling and volition that have become detached from their origins is entirely of a piece with the attention given to the ailing, abused body in Coleridge's travel letters and journals (its torments on this occasion exacerbated by opium withdrawal). It is the role of the body in determining the structure of the 'perceptual envelope' inhabited by the pedestrian traveller that needs finally to be amplified, since it is in the transparency of this operation that Coleridge's landscape descriptions in his notebooks are remarkable above all else. The reader is never allowed to forget the nuances of motion and perspective in the retarded seriality of these descriptions, as in the following extract from his 1799 tour of the Lakes:

> Grasmere [not the village, but the fell now known as Grasmoor] is a most sublime Crag, of a violet colour, patched here & there with islands of Heath plant – & wrinkled & guttered most picturesquely – contrast with the Hills on my Left Right, which tho' in form ridgy & precipitous, are yet smooth & green – We pass the Inn at Scale Hill, leaving it to our right & to our Right is Lowes Water which we see – tis a sweet Country that we see before us, Somersetshire Hills & many a neat scattered House with Trees round of the Estates Men. – The White Houses here beautiful – //& look at the river & its two-arched Bridges – We have passed it curved around the Hill – the Bridge, the Plain, & Lowes Water are at my Back – & before me – O God, what a scene. – the foreground a sloping wood, sloping down to the River & meadows, the serpent River beyond the River & the wood meadows terminated by Melbreak walled by the Melbreak.... close by my left hand a rocky woody Hill, & behind it, half hidden by it, the violet crag of Grasmere.... I climb up the woody Hill & here have gained the Crummock Water – but have lost the violet Crag.[31]

For a defining contrast, consider the account of the highway motorist's sensations in Appleyard, Lynch and Myer's study of the aesthetics of road construction, *The View from the Road*. The difference lies to a large extent in what the authors call the 'tempo of attention': for obvious reasons, a driver's vision is mainly directed forwards, and when travelling at a reasonable speed attention, which has a 'rushing, forced' quality, is 'concentrated on near objects straight ahead in the road'; as speed increases even more, the eye selects more distant objects, and the balance shifts 'from detail to generality'.[32] In Coleridge's narrative, by contrast, the progressional perspective is differently inflected: the tempo is very much slower, naturally, and attention is constantly shifting from the front to the rear, and to both sides, registering objects both near and far, and then registering them anew as the walker's situation and aspect alter. There is time to appeal to an implied reader ('look at the river & its two-arched Bridges') and to let the imagination play on particular sights (as in 'terminated by Melbreak walled by the Melbreak', where a fairly neutral phrase is immediately revised into something more figurative). The passage is rather typical of Coleridge's notebook entries in having a single, understated mention of what was evidently a more powerful visual or aesthetic experience: 'O God, what a scene'. Elsewhere a more striking effect is created by a sudden metaphorical excess out of keeping with the close, earnest descriptive technique that is generally practised: 'in the middle of the Lake the Screes commence far higher up, & occupy two thirds of the height in the shape of the apron of a sheet of falling water, <or a pointed Decanter, or tumbler turned upside down> or rather an outspread fan...'.[33] Whatever precise form these stronger impressions take, they are very economically deployed: if one adapts Appleyard, Lynch and Myer's idea of a roadscape having 'a basic beat, a regular frequency with which decisions and interesting visual impressions are presented', then the strong beats are probably too infrequent in Coleridge's travel writing to produce a rhythm that is either comfortable or compelling for the reader. But the fidelity with which this writing transcribes the experience of walking through a landscape is such an unprecedented Romantic development as to compensate for these defects.

Having taken stock of Coleridge's achievement as a writer of peripatetic in prose, I would like to devote the remainder of this chapter to assessing the extent to which the poetics of walking developed in his notebooks and letters finds a home within his poetry. One might expect that, given the social stigma historically attached to walking as a mode of travel, a rhetoric enlisted to the positive representation of walking might fall victim to self-censorship in work composed for publication, and that more normative aesthetic values would reassert themselves under the lengthening shadow of poetic tradition. On the other hand, that very tradition, as I showed in Chapter 3, had become more elastic in the course of the eighteenth century and, as we have seen with Wordsworth, was ready to cope with the cultural redefinition of walking as voluntary, recreational pedestrianism that took place towards its close. And Coleridge himself was willing to speculate on the most intimate correspondence between topography, pedestrianism and public verse, as the following notebook entry from 1803, delineating a project – a 'noble poem' – apparently akin to Wordsworth's *Prelude*, indicates:

> Go & build up a pile of three [stones], by that Coppice – measure the Strides from the Bridge where the water rushes down a rock in no mean cataract if the Rains should have swoln the River – & the Bridge itself hides a small cataract – from this Bridge measure the Strides to the Place, build the Stone heap, & write a Poem, thus beginning – From the Bridge &c repeat such a Song, of Milton, or Homer – so many Lines I will must find out, may be distinctly recited during a moderate healthy man's walk from the Bridge thither – or better perhaps from the other Bridge – so to this Heap of Stones – there turn in – and then describe the Scene – O surely I might make a noble Poem of all my Youth nay *of all my Life*....[34]

As Raimondo Modiano points out, Coleridge is here 'trying not just to evoke a certain place in a poem, but virtually to tailor his composition according to the physical pattern provided by the place':[35] although the entry is somewhat garbled, it seems that what Coleridge had in mind was a poem in which the number of lines was determined either, arbitrarily, by the number of strides measured between the two spots mentioned, or, in a slightly looser but equally contingent manner, by the time taken to walk between the two locations. It is hardly surprising that a poem conceived on such lines was never written, but the fact that Coleridge could envisage marrying physical and textual space in this manner is indicative of a commitment to 'writing the body' that extended beyond the intimate realm of his notebooks and private letters.

One of the formal and stylistic innovations in Coleridge's poetry to have been widely discussed, and much admired, is the blank verse of the Conversation Poems. 'Perspiration', the short poem written on the Welsh tour, is a forerunner of this style, developed in 'The Aolian Harp' the following year and brought to perfection, in many readers' eyes, in 'Frost at Midnight' in 1798. Many other Coleridgean poems (and all the successful ones) which have peripatetic themes, or which are rooted in his pedestrian way of life, share the same medium of loose, informal blank verse. There is, in fact, a remarkable caesura in Coleridge's poetic career that corresponds with the chronology of his life as an active walker: in the ten years between the Welsh tour and his departure for Malta in 1804, the years of his wanderings in the Quantocks and of his 'epic solo fell-walks' in the Lakes,[36] he produced around twenty-five significant compositions in blank verse; after his return from Malta, which effectively marked the end of his pedestrianism, I count only three finished poems in blank verse in the *Poetical Works*. I would hesitate to draw too mechanistic a link between these phenomena, but the coincidence is striking. I stand by my contention in Chapter 3 that the fluid, improvised quality of blank verse, its perceived 'naturalness' and its fundamentally syntagmatic ordering, made it the preferred metrical vehicle among literary pedestrians towards the end of the eighteenth century for the representation of a mode of travel celebrated above all for independence and freedom of movement. Coleridge's simultaneous retirement from pedestrianism and near-abandonment of blank verse makes this case more persuasive: where he explicitly structures his poem around progress through a landscape, blank verse appears the most compelling aesthetic choice; and even where walking does not figure thematically, it is often tempting to ascribe to Coleridge's blank verse the kinaesthetic properties so finely developed in his travel journals and prose landscape descriptions.

This iconic relation between the steady alternating rhythm of blank verse and the signified content of pedestrian travel is not necessarily disturbed by the presence within the poem of more traditional figurations of journey or pilgrimage. A good example is the 'Inscription for a Seat by the Road Side half-way up a Steep Hill facing South', written in 1800. This poem, apparently written to supplement Wordsworth's 'Poems on the Naming of Places' in the second volume of *Lyrical Ballads*, is addressed to youthful passers-by who might be brought to reflect on the less fortunate: the 'foot-worn soldier and his family', the migrant labourer returning home, or those

> Who in the spring to meet the warmer sun
> Crawl up this steep hill-side, that needlessly
> Bends double their weak frames, already bowed
> By age or malady, and when, at last,
> They gain this wished-for turf, this seat of sods,
> Repose – and, well-admonished, ponder here
> On final rest.   (ll. 20–5)[37]

It is at this point, when the 'seat' as a place of rest becomes an occasion for meditating on 'final rest', that the poem is overtaken by conventional religious allegory of life-as-journey. The carefree young traveller is now imagined as being morally invigorated by these reflections to construct a 'seat' on which his soul may find eternal refreshment:

> a seat,
> Not built by hands, on which thy inner part,
> Imperishable, many a grievous hour,
> Or bleak or sultry may repose – yea, sleep
> The sleep of Death, and dream of blissful worlds,
> Then wake in Heaven, and find the dream all true.   (ll. 36–41)

In line with the conventional strategy of the 'inscription' genre, the poem's whole intention is to arrest the traveller-reader, in this case with the sudden deepening of topographical significance; but the attenuated, broken rhythms of the first half of the poem, which seem to accompany well the physical expenditure depicted, create too powerful a 'set' to be undone by this interruption at the level of the poetic argument, creating a tension between formal and semantic elements: the poem continues its 'o'erlaboured' progress in despite of the implied reader's meditative pause.

Comparison might be made here with the contemporaneous 'A Stranger Minstrel', addressed to Mary Robinson. This is another poem of arrested travel, which places the speaker half-way up another hill (or rather, mountain), Skiddaw, where sweet/sad thoughts of Mary Robinson, who is in the throes of a terminal illness, elicits a reproof from the mountain to the effect that Coleridge's plangent desire for her company shows insufficient respect for her 'unfettered' soul. A sharp contrast is therefore drawn between the poet as 'supine' walker and the 'magic song' of Robinson, which takes its natural metaphor from aerial transport:

> No wind that hurries o'er my height
> Can travel with so swift a flight.   (ll. 59–60)

Since Coleridge is supine at the start of the poem, the sense of physical and mental arrest cannot take purchase in the same way that it does in the 'Inscription'. The possible implied counterdefinition of Robinson's aerial imagination and Coleridge's more pedestrian muse also receives little assistance from the poem's formal structures, since the mixed decasyllabic and octosyllabic couplets incline it towards an incantatory format: the periodic closure which the couplet form enacts allows for no effect of progressionality. Another interesting comparison is with the earlier 'To a Young Friend' (1796), homage to the poet's ill-fated relationship with Charles Lloyd. This poem begins by offering, in place of some anonymous mountain, 'wearisome and bare and steep', the picturesque roughness and verdant complexity of

> a green mountain variously up-piled,
> Where o'er the jutting rocks soft mosses creep,
> Or colour'd lichens with slow oozing weep....   (ll. 2–4)

Coleridge pictures himself and Lloyd alternately resting pensively in one of the mountain's green recesses, and continuing their progress 'up the path sublime' (l. 17), relishing the ever-widening prospect. The emphasis, sustained through dense patches of nature-description, is on the companionable nature of the walk, on the equal intimacy and freedom of peripatetic conversation, study and creativity:

> In social silence now, and now to unlock
> The treasur'd heart; arm linked in friendly arm,
> Save if the one, his muse's witching charm
> Muttering brow-bent, at unwatch'd distance lag;
> Till high o'er head his beckoning friend appears,
> And from the forehead of the topmost crag
> Shouts eagerly....   (ll. 25–31)

'To a Young Friend' is written in irregularly rhyming pentameter, which allows for the intonational contour and syntactic coherence to be projected across line-breaks, availing the poem of most of the freedom of blank verse proper and reinforcing the sense of irregular yet determined progression that is conveyed at the semantic level.

Coleridge's description of the 'topmost crag' as a place 'to cheat our noons in moralising mood' (l. 43), as enabling philosophical clear-sightedness removed from 'the world's vain turmoil' (l. 39), is the only tangible preparation for the half-hearted allegorisation of the ascent which then follows. This seemingly gratuitous generic gear-change startles with its brisk re-assignment of symbolic significance to what had previously appeared as naturalistic topographical details: thus the green mountain becomes the 'Hill of Knowledge' (l. 50), the 'unsunn'd cleft' (l. 36) re-emerges as the hiding-place of 'Inspiration' (l. 56), and the 'shadowing Pine' is now seen to have obscurely suggested the 'evergreen' strength of a mind superior to 'Want', age and the trials of 'Bigotry' (ll. 58–60). However, this symbolic recuperation of the walk is, to this reader at least, only partially achieved: the naturalism of Coleridge's descriptions resists the 'rude' intrusion of 'allegoric lore' (l. 49), and the poem concludes by returning the two walkers to a vantage-point that invites literal interpretation:

> There, while the prospect through the gazing eye
>   Pours all its healthful greenness on the soul,
> We'll smile at wealth, and learn to smile at fame,
> Our hopes, our knowledge, and our joys the same,
>   As neighbouring fountains image each the whole:
> Then when the mind hath drunk its fill of truth
>   We'll discipline the heart to pure delight,
> Rekindling sober joy's domestic flame.   (ll. 67–74)

Here it would seem perverse to construe the 'truth' which the mind drinks as a mental abstract: it seems inseparable from the 'healthful greenness' that enters via the eye. In a typically selfless, even sacrificial, act of imagination, Coleridge ensures that this greenness refreshes the souls of both men indifferently, their social walk having secured, in Coleridge's trusting anticipation, a brotherly fellowship of natural wisdom.

The contrast, in 'To a Young Friend', between an exposed hilltop commanding extensive views and a sheltering recess or lonely habitation favouring contemplation and creativity, is a familiar one in Coleridge's conversation poems. It is a contrast of positives which invites interpretation in terms of the biological landscape aesthetics elaborated by Jay Appleton. For Appleton, the sophisticated categories of the connoisseur of natural beauty are ultimately reducible to more primitive behavioural responses:

> [A]esthetic satisfaction, experienced in the contemplation of landscape, stems from the spontaneous perception of landscape features which, in their shapes, colours, spatial arrangements and other visible attributes, act as sign-stimuli indicative of environmental conditions favourable to survival, whether they really *are* favourable or not.[38]

Appleton does not, of course, argue that such landscape-response is normally a self-conscious process; it is rather a matter of unconscious programming, arising from the sublimated operation of survival mechanisms that have ceased 'to be needed for some practical, utilitarian purpose'.[39] He narrows his focus on environmental stimuli still further to concentrate on the potential offered by a particular habitat for seeing without being seen – thus yielding 'prospect-refuge theory'. The typical topographical structure of a Coleridgean conversation poem (followed, for example, by 'This Lime-Tree Bower My Prison', 'The Aolian Harp', 'Reflections on Having Left a Place of Retirement' and 'Fears in Solitude') is to move from a refuge or refuge-dominant landscape to a prospect setting, which may bring about an enlargement of the poem's conceptual as well as spatial horizons, and then to return to the original refuge at the close. 'To a Young Friend' interestingly turns this pattern inside-out – as does, in miniature, 'Lines Composed While Climbing the Left Ascent of Brockley Coomb' (1795).

This ecological interpretation of Coleridge's aesthetic preferences could be complicated by introducing a third environmental factor (and consequent natural symbol) which Appleton subordinates in his account, that of the 'hazard'. Hazards are those properties of the landscape which affect the observer's ability to escape from danger, and include a category of 'locomotion hazards' threatening one's freedom to move swiftly across country to a place of refuge. It is difficult to reconcile the idea of a phylogenetic affinity for open or navigable space with Coleridge's stated preference (as a pedestrian) for crowded landscapes and impeded progress. I have already, in the course of my discussion of Coleridge's prose, linked this preference to the metaphysical anxieties accruing to unprotected space as anatomised by Michael G. Cooke. Among the many excellent examples treated by Cooke one might briefly rehearse the memorable wasteland opened up in Coleridge's fragmentary prose poem, 'The Wanderings of Cain' (1798):

> The scene around was desolate; as far as the eye could reach it was desolate: the bare rocks faced each other, and left a long and wide interval of thin white sand. You might wander on and look round and round, and peep into the crevices of the rocks and discover nothing that acknowledged the influence of the seasons. (ll. 70–5)

Here the rock crevices betray their promise as refuge symbols by refusing to support life; in a similar way, it is under a larger rock, where his son Enos had previously found a pitcher of water, that Cain encounters the ghost of his past sins, 'the Shape that was Abel' (l. 126). Equally, the landscape's bleak prospect symbols, the 'pointed and shattered summits of the ridges' that seemed to resemble 'steeples, and battlements, and ships with naked masts', are no more than 'a rude mimicry of human concerns' (ll. 81–4), withholding the security, companionship and purpose which the metaphors cruelly betoken. Coleridge's description of the wilderness cancels out whatever signs appear to favour survival: even the absence of locomotion hazards is vastly overcompensated by the 'deficiency hazards' (Appleton's term) associated with desert terrain. Cain's passage across the land is grimly devoid of destination or purpose: walking is here, as in many of Wordsworth's poems of the 1790s, the bodily displacement of spiritual suffering and exile. When the three figures depart, it is with the profoundly negative goal of discovering 'the God of the dead', with little expectation that Cain's desperate existential queries will be answered.

It is ironic that Cain yearns for 'darkness, and blackness, and an empty space' (l. 289); but as Cooke suggests, the desolation he craves borders on complete insensibility, and this perversion of desire does not derail Coleridge's primary strategy of 'using space morally, and humanistically, to portray a man's influence, or lack of influence, over his own destiny'.[40] 'The Wanderings of

Cain' provides support for Steven Bourassa's argument in favour of supplementing Appleton's relentless biologism with attention to the cultural determination of aesthetic preferences. Such preferences, Bourassa maintains, 'signify values that stabilize cultural, group, or individual identity',[41] and presumably perform this function whether the particular symbolic meanings are positively or negatively marked. Whatever quasi-instinctual grounds there might be for valuing open, easily traversable ground, Coleridge's Christian learning and allegiance ensure that in his poetry such landscapes are invariably overwritten with the symbolism of spiritual desert and with ideas of moral errancy. He is no more able than Cowper to disambiguate the approbatory (free, self-fulfilling) and pejorative (lost, self-alienated) senses of 'wandering'.

Within the significant corpus of Coleridge's 'journey poetry' (in Raimondo Modiano's phrase), or that larger part of it which figures a journey or excursion as pedestrian passage, the poems which seem least encumbered with metaphysical disquiet, and least inhibited by the topographical convention of tracing moral analogies between the natural world and human affairs, are a small handful of blank verse compositions – the most striking and successful being 'This Lime-Tree Bower My Prison'. The early 'Reflections on Having Left a Place of Retirement' (1795) shows Coleridge struggling to reconcile the literal representation of rural walking with the spiritual topography of hill-top vision, and to reconcile both these with the rhetoric of religious crusade ('I therefore go...to fight the bloodless fight' [ll. 60–1]). The poem begins in the type of pastoral domestic 'refuge' that forms the emotional base of several of the conversation poems: a 'pretty Cot' (l. 1) surrounded with myrtle and jasmin, apparently the sole human settlement in a 'Valley of Seclusion' (l. 9). From there the scene switches to a 'stony Mount' (l. 27) from which Coleridge surveys a various landscape and sees in it divine 'Omnipresence' (l. 38). Then he seemingly brackets cot and mount together in severely overcompensatory reflections on the moral deficiencies of rural retirement, before making his resolve to fight the bloodless fight for a world in which all are possessed of the blessings he has enjoyed.

The transitions between these different sections of the poems are rather awkwardly managed: the reader lurches from 'refuge' to 'prospect' in the middle of line 26, and is then bemused in the third verse paragraph by a self-inquisition that implicitly discredits the positive experiences previously described. In terms of its spatial horizons, Coleridge indeed seems, as Michael Cooke suggests, to vacillate 'between claustro- and agoraphobia',[42] and can enthusiastically embrace neither retirement nor mobility. Pedestrian experience is validated most fulsomely in the second verse paragraph, which begins by noting the 'perilous toil' required to reach the top of the stony Mount where Coleridge achieves a state of sensory and spiritual plenitude:

> Oh! what a goodly scene! *Here* the bleak mount,
> The bare bleak mountain speckled thin with sheep;
> Grey clouds, that shadowing spot the sunny fields;
> And river, now with bushy rocks o'erbrow'd,
> Now winding bright and full, with naked banks;
> And seats, and lawns, the Abbey and the wood,
> And cots, and hamlets, and faint city-spire;
> The Channel *there*, the Islands and white sails,
> Dim coasts, and cloud-like hills, and shoreless Ocean –
> It seem'd like Omnipresence!   (ll. 29–38)

Kelvin Everest argues that the description of Coleridge's 'increasingly difficult, but increasingly excited ascent', with its perspective 'widening and lengthening' as he gets higher, is made to stand for the 'strenuous effort necessary to achieve a sense of the full potential in nature' and 'its capacity of manifest the presence of God'.[43] I think this overestimates the somewhat perfunctory account of the ascent, and misreads as perspectival variations what are given as different components of the same hilltop panorama, but the spirit of Everest's reading is a sympathetic one. Insofar as it obviously catalogues objects from a wider sweep of land and sea than could be comprehended in a single gaze, and in the way that the movement of that gaze is from *here* to *there* (rather than being snatched immediately by the luminous distance, as in a typical Claude landscape), the passage strikes me as anti-picturesque; and by the same naturalistic token the physical effort and visual awareness of the walker are confirmed and legitimised.

The key elements of 'Reflections' – refuge and prospect, private contentment versus social guilt – are found in more spectacular disjunction in 'Fears in Solitude', the most explicitly politicised of the conversation poems. At the beginning of the poem the speaker's walk has deposited him in 'A green and silent spot, amid the hills' (l. 1), which invites a calmer scrutiny of 'Religious meanings' (l. 24) in natural forms than the hilltop vision characteristically provides. The pastoral situation of utterance is then virtually eclipsed for nearly two hundred lines, as the encounters of the walk are displaced by increasingly rancorous rhetorical encounters with the political and religious demons of the historical moment. The patriotic fears which accompany this *jihad* on British immorality, corruption and bloodlust encourage a renewed appreciation of the beauties of his native landscape – a natural temple in which, Coleridge writes in Cowperesque phrase, 'I walk with awe' (l. 196).

The walk back to Stowey finds Coleridge 'Startled' (the emphatic caesura following this word mimics the arrested motion) by another coastal prospect, its 'mighty majesty' benignly accommodated to human needs and projective capacities, and serving the poet as a metaphor for the harmonious community that he despairs of finding in actuality:

> On the green sheep-track, up the heathy hill,
> Homeward I wind my way; and lo! recalled
> From bodings that have well-nigh wearied me,
> I find myself upon the brow, and pause
> Startled! And after lonely sojourning
> In such a quiet and surrounded nook,
> This burst of prospect, here the shadowy main,
> Dim-tinted, there the mighty majesty
> Of that huge amphitheatre of rich
> And elmy fields, seems like society –
> Conversing with the mind, and giving it
> A livelier impulse and a dance of thought!   (ll. 209–20)

Coleridge then picks out Wordsworth's Alfoxden 'mansion' (l. 223) and his own 'lowly cottage' (l. 225), and the earlier invocation of the 'bonds of natural love' (l. 180) now gathers its personal referents. His contemplative repose has excited, rather than suppressed, 'the thoughts that yearn for human kind', even though family and close friends are the only immediate objects for these affections. In 'Fears in Solitude', by contrast with 'Reflections', natural objects and extended familial life together stand as the social alternative to the wise passiveness of rural retirement. This defines a more modest rhetorical function for the walk in this poem than for the crusading march in 'Reflections': Coleridge's excursion frames his political sermon and the enunciation of his patriotic fears, and the walk back aims at connecting the public themes articulated in the valley of seclusion to a world beyond that is, ironically, as private and bounded as the one he has left.

In 'This Lime-Tree Bower My Prison', by suppressing all evidences of his social mission and accepting the limitations of his conversational situation of utterance, Coleridge writes his finest piece of peripatetic verse. Despite its origins in an accident that 'disabled [him] from walking', the poem is emphatically one about walking. Moreover, it is a poem which exemplifies a number of the characteristics of pedestrian practice, in particular the pedestrian's experience of landscape: a sensuous, participatory relation with the environment; a moderate, steady rate of passage that provides freedom to stop and restart at will; a sense of human proportion vis-a-vis the natural environment, hence an 'ecological' consciousness sharply opposed to the sense of mastery of space enjoyed by the horse-rider or the modern motorist; and a progressional, rather than logical or conceptual, ordering of reality, a mental fidelity to the temporal and spatial sequences in which the world was experienced.

'This Lime-Tree Bower' has received extensive critical attention for its treatment of the speaker's relationship with nature and its performance of an empathetic imagination, but the significance of its representation of passage, albeit an imagined passage, through a landscape has been less often noted. Kelvin Everest, in his detailed commentary on the poem, observes how the expansion of consciousness is 'co-extensive with the progressive widening of [physical] horizons on his imagined journey'; and Raimondo Modiano, in highlighting the importance of different forms of confrontation or encounter (between man and nature, between domesticity and public life, and so on) in the Conversation Poems generally, includes 'This Lime-Tree Bower' under the heading of Coleridge's 'journey poetry'. In a particularly interesting reading, Anne Mellor diverts attention away from the poem's mimesis of an actual walk onto what she presents as an aesthetic progress through the categories of the picturesque, the beautiful and the sublime – though one that ends by collapsing these categories into a moral and spiritual responsiveness to nature that overcomes the boundary between spectator and aesthetic object.[44]

I do not believe that these approaches are irreconcilable, but rather take the view that the aesthetic progress Mellor describes is co-ordinated with the imaginative reconstruction of material passage through an individualised landscape. I would argue furthermore that the specific modalities of pedestrian passage are not only legible within the text but exercise an influence on its larger argument. To begin with, the imagined journey of Lamb and the Wordsworths is one that respects (however artfully) the unhurried pace, the effort, and the irregular line of the walk. One is put in mind of Coleridge's own pedestrian habits and preferences as he traces the path of his friends who 'On springy heath... wander,... wind down... To that still roaring dell,... wander on / In gladness'. The reference to Lamb 'winning [his] way... through evil and pain / And strange calamity' almost seems to be arrived at by metaphorical appropriation of the idea of difficult passage over rough terrain. The complicated syntax of lines 5–20 (a single sentence), which combines the broken rhythms of a line like 'The roaring dell, o'erwooded, narrow, deep' with the greater ease of lines like the next one, a single rhythmic unit ('And only speckled by the mid-day sun') is suggestive of uneven progress through a landscape which, in picturesque fashion, offers locomotive as well as visual obstructions. The sensuous engagement with the surroundings, and the particularity of the observations, both of which pertain to the walker's experience of landscape, need no elaboration; while the repetitive, overlapping movement that is a striking feature of these lines (the doublings on the 'roaring dell', the ash, 'tremble' and 'drip') convey the pauses for a more lingering gaze that are made possible by the pedestrian's freedom of movement.

The stationary contemplation of the sublime sunset (a positive sublime, as Anne Mellor points out, from which fear and self-humbling are absent), and the resumption of attention to the bower in which the speaker is imprisoned, are evidently moments in which the surrogate walker's passage is arrested. And yet the serial movement that has characterised the first half of the poem is grammatically replicated in the co-ordination of Coleridge's detailed observations of foliage and sunlight via simple connectives, and echoed thematically in the appeal to what Anne Mellor calls the 'never-ending series of positive emotional experiences' which 'living nature' promises the 'responsive viewer'.[45] The poem's concluding image, which provides one final projective link between Coleridge and Lamb, fascinatingly asserts this human bond in a figure of passage:

> My gentle-hearted Charles! when the last rook
> Beat its straight path along the dusky air
> Homewards, I blest it! deeming its black wing
> (Now a dim speck, now vanishing in light)
> Had cross'd the mighty Orb's dilated glory,
> While thou stoods't gazing; or, when all was still,
> Flew creeking o'er thy head, and had a charm
> For thee, my gentle-hearted Charles, to whom
> No sound is dissonant which tells of Life.   (ll. 68–76)

Here, the progressional perspective that was established in the early parts of the poem is seen controlling the expression of its emotional and spiritual theme: the rook, passing out of the gaze of one and into the gaze of the other, rhetorically unites the friends in a shared act of worship. The rook's crossing of the circle of the sun stands as a summary of the poem's method, which disregards the circular logic of the symbol in favour of a logic of sequence, and which performs a crossing of boundaries (between self and nature, between past, present, and future) without transcending them. It is ironic that in the straight path which it beats across the sky the rook is a poor counterpart of the poet who throve on the resistance of uneven ground, and who delighted in parting from the straight high road of literary language to cut across the pathless field of novelty.

Written less than two years after 'This Lime-Tree Bower', the 'Lines Written in the Album at Elbingerode, in the Hartz Forest' (dated 17 May 1799) indicate that Coleridge's enthusiasm for nature was a more unstable and contingent passion than might have been supposed from a reading of the conversation poems. Containing, according to Coleridge, 'a true account of my journey from the Brocken to Elbinrode' on the Harz walking tour of May 1799,[46] the poem presents the weary speaker as having exhausted his responsiveness to the energy of natural forms: his dragging feet (bladdered and swollen, according to his letter to Sara) mark his separation from the *surging* scene' of 'Woods crowding upon woods, hills over hills' (ll. 2–3), and he seems as discrete an element of this scene as the various 'most distinct' (l. 10) sounds – birdsong, breeze, brook, goat bell – that he dutifully records. There is no happy congruence between the pedestrian's freedom of movement and the restless picturesque gaze; rather,

> the sight vainly wanders nor beholds
> One spot, with which the Heart associates
> Holy Remembrances of Child or Friend,
> Or gentle Maid, our first & early Love,
> Or Father, or the venerable Name
> Of our adored Country.   (ll. 18–23)

The aesthetic appreciation of nature available to the settled man perambulating his local landscape in the conversation poems is denied to the de-territorialised speaker of this poem, who finds that human associations are more important than pictorial ones. Equally, the blissful sense of divine 'Omnipresence' which Coleridge celebrated in 'Reflections on Having Left a Place of Retirement' is here reluctantly surrendered to the 'sublimer Spirit' (l. 35) of one more spiritually self-assured; for the homesick Coleridge, God plays second fiddle to his 'delegated Deity', England (ll. 23–5). The most remarkable feature of this poem is the way in which what one critic, seeking continuities between Romanticism and contemporary preoccupations, has described as Coleridge's 'cosmic ecology',[47] is seen to be dependent on contingencies of place and habitation.

As Raimondo Modiano has helpfully shown, 'Lines Written in the Album at Elbingerode' shows Coleridge 'moving towards a poetry that will emphasize the "finer influence from the Life within" and resist the influence of "outward forms"'.[48] As a consequence of this shift, the materiality of the Romantic pedestrian is lost in favour of symbolic re-appropriations of walking and wandering. The best illustration of this tendency, and a fitting poem to conclude this study of Coleridge, is 'The Picture' (1802).

This poem begins with a seemingly realistic description of a walk over rough terrain, coupling a typically Coleridgean pleasure in obstructed passage with a joyful indeterminacy of goal reminiscent of the opening to Wordsworth's *Prelude*:

> Through weeds and thorns, and matted underwood
> I force my way; now climb, and now descend
> O'er rocks, or bare or mossy, with wild foot
> Crushing the purple whorts; while oft unseen,
> Hurrying along the drifted forest-leaves,
> The scared snake rustles. Onward still I toil,

> I know not, ask not whither! A new joy,
> Lovely as light, sudden as summer gust,
> And gladsome as the first-born of the spring,
> Beckons me on.... (ll. 1–10)

The joy is said to depend partly on having quelled the 'master-passion' of love; but, as the poem progresses, it becomes plain that the narrator's triumph over this master-passion was prematurely announced. He tries to oppose his own sceptical intellect to the obsessional fantasies of a 'lunatic' lover, but the boundary between the two personae becomes ever more insubstantial. Throughout, the oscillation between realism and idealism is articulated as the stages of a walk. In a manner analogous to the alternating mental rhythm of pedestrian travel – to the shifting of attention from external surroundings to one's own thoughts, and back and forth between the two at variable intervals – the sceptical phases take the form of close encounters with sensuous nature, whereas the reascendancy of the master-passion is marked by moments of arrested motion at which contemplation can loiter.

In the first stage of his forest walk the narrator manifests a close tactile awareness of the ground over which he travels with difficulty; but his observation that the trees on either side converge to form a 'melancholy vault' (l. 15) above him initiates a digression on a 'love-lorn man' (l. 18) whose sentimental imagination falsely invests the elements with supernatural life, and who is apt to be comically disabused when his 'dainty feet' encounter 'the briar and the thorn' (l. 30). Laughing off such delusions, the speaker moves on to find another notably unidealised object, a 'Hollow, and weedy, oak' (l. 50), but no sooner does he ensconce himself in its shade than a fresh reverie about the foolish lover begins. His antiromantic assertions that the breeze 'Was never Love's accomplice' (l. 59) and 'never half disclosed / The maiden's snowy bosom' (ll. 62–3), and that the neighbouring stream never reflected the 'stately virgin's' (l. 74) heavenly body, ironically lead him to a full-blown identification with the fantasy he is ostensibly repudiating. The imperative 'Behold!' (l. 76) positions the speaker as perceiving consciousness, co-worshipper of the 'watery idol' (l. 83), and for more than thirty lines of excited description the 'sickly thoughts' he admonishes seem as much his as they are those of the third-person lover. To rouse himself from this mental paralysis, he once more resumes his 'devious course' through the forest, but his desire for 'deeper shades and lonelier glooms' (ll. 120–1) is not a promising sign, and very soon he is re-immersed in erotic misreadings of the landscape, without even a token externalisation of the experience in a fictional alter-ego: the division and re-unification of a stream is presented uncritically as a metaphor of sexual union, 'each in the other lost and found' (l. 127).[49]

From this point the love-distempered youth drops out of sight, or rather merges with the identity of the speaker. Resuming his walk no longer has a curative effect upon the latter: the next object he passes, a birch tree, is assimilated to his melancholy ('a *weeping* birch' [l. 136; my italics]) and personified as female ('the Lady of the Woods' [137]). In the final stage of his walk, the narrator enters upon a picturesque landscape of crescent hills, stone cottages and fruit trees, and discovers a representation of the same scene sketched in berry-juice on a piece of tree bark. He instantly construes this as the work of his beloved Isabel, and, overcoming the temptation to dwell in morbid satisfaction on his 'relique', decides to pursue and assist her on her hazardous journey through the 'long wood' to her father's house. I cannot agree with Raimondo Modiano, who gives an otherwise very illuminating reading of 'The Picture',[50] that this represents a healthy resolution of the speaker's vacillation between down-to-earth empiricism and self-consuming romantic fantasy. I side rather with H. R. Rookmaaker, who argues that 'nature has become a screen on which [the protagnist's] feelings are projected without the sobering intervention of reason informing him of the subjective character of his vision'.[51] There are no apparent grounds for his conviction that the sketch is the work of Isabel, and the depressed 'patch of heath' that he drools over as her recent 'couch' (l. 165) could equally well have been formed by sheep, as in lines 36–9. Moreover, he cannot know that she is 'alone', as he exclaims in line 178; he *wants* her to be alone, so that he can become her devoted guide through the forest, and his beliefs again come into line with his wishes. His decision to 'restore' rather than 'keep' his 'relique', far from demonstrating his freedom from false idolatry, merely commits him to a wild goose chase propelled by the sort of distempered fancies he began by deriding. As such it is a critical waymark to the preoccupations of Coleridge's later poetry, in which the walker's commitment to the *différance* of life is replaced by constancy to a range of ideal objects. As the most adventurous of all Romantic pedestrians, Coleridge vigorously sought out the multiplicity of natural forms and appearances, bringing them before a sensibility that could encompass picturesque curiosity, at its most playful, and a visionary reading of natural signs as the language of God, at its most elevated; as his years of travel and pedestrianism come to an end, his favourite image of a mountain walker seduced by the Brocken spectre (his own shadow projected on the morning mist) comes to stand for his revised belief that no degree of mobility and change, no degree of repetition of the anxieties and pleasures of arrival and departure, can ultimately do much to liberate one from the omnipresence of self.[52]

6
Gender, Class and Walking: Dorothy Wordsworth and John Clare

In the early Romantic period, as we have seen, walking rapidly assumed the character of a voluntary, pleasurable activity: travel without the *travail* it had traditionally connoted. The loss of those socio-cultural significances, or the newly ironic manipulation of them, were indispensable to the emergence of a new form of masculine, middle-class self-fashioning. The vast majority of the early pedestrians I surveyed in the early chapters of this book fall within this latter categorisation, and William Wordsworth and Samuel T. Coleridge, on whom I have focused more attention, confirm the bias: both, though barely solvent at times during the 1790s, were respectable and expensively educated, confident of the privileges of their class and gender, and socially conditioned to seek intellectual rather than bodily labour. But what of those outside the ranks of the culturally empowered, for whom travel in general was not a natural prerogative, for whom geographical mobility was not the ready expression of their personal freedom, and for whom the choice of pedestrian transport did not, could not, perform the same values as it did for middle-class men? Labouring-class men, and women of whatever class, inhabited material contexts which impeded their participation in the age of pedestrianism, and where they nevertheless achieved distinction as writers through their walking, as did the two figures named in the title of the present chapter, it was against the pressure of very different ideological constraints.

A brief description of two remarkable journeys – one quite familiar, the other very little-known – will help focus some of the issues to be discussed below.

In the spring of 1822, Sarah Hazlitt arrived in Edinburgh to initiate divorce proceedings against her estranged husband William, then suffering a crisis in his notorious relationship with his landlord's daughter, Sarah Walker. The divorce was of his seeking, but Sarah understandably wished to effect a favourable settlement for herself. In the course of the next three months, as she waited for the legal machinery to complete its work, she undertook two unaccompanied pedestrian tours (in addition to a trip to Ireland), which Hazlitt was later to object to financing. She calculates the first of these trips as comprising a total of 170 miles, with the longest day covering an impressive 32 miles.[1] Throughout her time in Scotland she maintained a journal, which is an intriguing mixture of travel writing and diaristic accounts of her daily life and of her dealings with Scottish lawyers and Hazlitt himself.

Sarah Hazlitt's writing is often humdrum and derivative, but is of considerable interest in giving us a rare personal record of a notable female pedestrian. As such, one inevitably looks for gender-markings in her representation of the experience of travel. Her close interest in practical life and domestic arrangements are an obvious form of testimony: her second tour, for example, affords a realistic description of a poor Highlander's home:

> there is in general but one room for all the family, and the beds, are presses or cribs made in the wall, so that there is no appearance of any, there is scarcely any furniture, but two or three old broken stools and chairs, with a pot to boil the victuals, a meal tub, and a few wooden spoons and tubs for eating; the women and children have neither caps shoes or stockings. Yet they seem strong, hearty, and comfortable.[2]

The care with which she transcribes the conversation of the Highlanders she encounters shows an ear for dialectal variation that goes beyond the picturesque or naive-primitivist othering of Scotland and the Scots found in some travellers' writings. Equally, her attention to the physical hardships of pedestrian travel constantly short-circuits any aesthetic pretensions her narrative assumes. Arriving at Loch Katrine via the Pass of Leney, her eye is diverted from 'the magnificence of the scenery' to the utility of a road cut into the rock along the side of the lake, which prevents the inhabitants having to clamber 'along the face of the precipice by the help of the roots and branches of trees'. Walking to Ben Lomond across 'a wide and dreary moor, full of bogs', she gets caught in a storm and the fear of losing herself 'almost overcame' her. After a 17-mile walk from Lanark to West Calder across another boggy moor, she sits down several times 'from mere inability to proceed', nursing her swollen and painful feet, and the following day she writes graphically of the material pleasures (such as the possibility of a 'thorough ablution') of her return to Edinburgh.[3] These passages illustrate Sarah's projection of the kind of embodied selfhood that recent feminist critics have seen as more characteristic of female Romantics. A down-to-earthness of another kind appears in places where she implicitly counters the rhetorical exaggeration by many tour-writers of the dangers of rugged terrain. At the Bracklinn Bridge near Callander, a narrow, unfenced structure over a ravine, she quotes Scott's remark that it is 'scarcely to be crossed by a stranger without awe and apprehension', then adds disarmingly: 'Nevertheless, I crossed, and explored every part ....'[4]

One perhaps surprising aspect of Sarah's account is her discovery that 'You may walk all through the country without molestation or insult'.[5] Although this assertion implicitly acknowledges the fear of sexual harassment or assault, and although some of the people she encounters have difficulty understanding her motives for travelling alone, the hospitality they afford her seems unaffected by suspicions that she might be a pedlar or 'a ladies maid going to my place',[6] and her passage is both safe and companionable. This strengthens the positive tone of the 'Journal', which conveys a delight in her own physical agility and fortitude (along with the discomforts and indignities), a zest for experience encompassing both human encounters and sensuous-aesthetic responses to natural scenery, and (very occasionally) a more reflective, spiritual capacity. The needs of body and spirit are memorably entwined in the following description of her passage through Glen Amon on the way to Crieff; her testimony to religious thoughts gives a Biblical resonance to the image of drinking water, which at first is arrestingly corporeal:

> After leaving it [Glen Amon], I had still other mountains to climb in succession; a most laborious road, and I should have been utterly exhausted with fatigue and heat, had I not found some mountain springs in my way; and lay down and bathed my face, and drank to allay the parching thirst. I was but clumsy at it at first, but I soon managed so as to drink very well, and was refreshed , and was thankful that God had provided water in the stony rock. These walks always make me more religious and

more happy, more sensibly alive to the benevolence and love of the Creator, than any books or church.[7]

Sarah Hazlitt's journal is indeed a rare document: outside the work of Dorothy Wordsworth, and with the minor exception of Mary Shelley's *History of a Six Weeks' Tour*,[8] I can find no substantial account of a pedestrian tour in the Romantic period undertaken by a woman. Mary Wollstonecraft, whose *Letters Written During a Short Residence in Sweden, Norway, and Denmark* has a deservedly high status in Romantic travel writing, was not fundamentally a pedestrian traveller. Nor, for example, were Helen Maria Williams or Ann Radcliffe, who also made interesting contributions to the genre. Jane Robinson's very helpful descriptive bibliography, *Wayward Women*,[9] offers little new evidence of female pedestrianism in the period, although it is clear from passing references in other tours that such activity was increasing.

The paucity of literature of this type is, of course, a reflection of the relative scarcity of women's travel writing generally in the late eighteenth-/early nineteenth centuries, which in turn reflects the historical, deep-rooted gendering of travel itself as a masculine activity. James Clifford argues that the very term 'travel' is bound up with 'European, literary, male, bourgeois, scientific, heroic, recreational, meanings and practices', and that although there were more early women travellers than was once realised, they 'were forced to conform, masquerade, or rebel discreetly within a set of normatively male definitions and experiences'.[10] Shirley Foster's study of Victorian women travellers would seem to bear out this theory, finding in their written accounts a gender ambiguity that honours both the 'masculine' virtues they necessarily cultivated as travellers and the cultural models of true femininity which still exerted pressure on female self-representations in literature.[11] It is part of the interest of Sarah Hazlitt's journal that it constantly flouts the conventions of masculine 'colonial' discourse, which Foster sees as the counterinfluence in female travel writing to conventions deriving from women's fiction (delicacy, sensibility, and so forth).

If travel in this period is seen as dangerously masculinising for women, it is because of the transcultural historical reality of the mobility of men versus the territorial stability of women, a stability that has served purposes primarily of safe reproduction.

Eric Leed cites observations by the anthropologist Nancy Munn on gender roles in Gawan culture that have a much broader application:

> In Gawa, women are associated with immobility, permanence, heaviness, soil, renewals of time, gardens, interiors, bounded space, security, and lack of freedom. Men are associated with exteriors, unbounded space, danger and insecurity, light things – sea and air – and abnegations of time in speed....[12]

The resulting 'genderization of space' is, as Leed says, one that persists in modern times, when travel has become accessible to individuals of either sex if they have the required leisure and money, without erasing absolutely the inherited associations of women with place, domesticity, and bounded lifestyles. If the presupposition of such a 'natural' binding of desire still has purchase today, it must have been very much stronger in the period under discussion, a pre-industrial tourist era in which women had only recently begun to travel in greater numbers (typically as members of a family group) as part of the middle classes' appropriation of the Grand Tour in the second half of the eighteenth century. And if travel of whatever kind threatened to desex or compromise the reputation of the respectable woman – travel for elopement or to live down a scandal were common, and conservative disapproval of female travel was increased by a belief in the loosening of sexual virtue that took place 'on the road'[13] – travelling on foot, with its socially levelling connotations, could only aggravate the fault. Within this environment of assumptions and values, Sarah Hazlitt's solitary tours appear courageously nonconformist.

Whatever the inhibitions, dangers and possible social penalties for a middle-class woman, the lower classes, women and men alike, had always walked. This walking was typically of a local, repetitious, involuntary and functional character, often connected with the search for, or daily practice of, work, although recreational walking, especially on Sundays, was by no means uncommon. The second journey I would like briefly to examine, although in important respects unrepresentative of its subject, is one that throws a vivid light on these very different conditions of passage.

In July 1841, John Clare 'escaped' from the High Beech asylum in Epping Forest and walked home to Northborough, taking four days to complete the journey. It is suggestive that, as Roy Porter points out, no one *en route* seems to have taken him for a lunatic,[14] and his ironic use of military metaphors to describe his escape ('Reconnitered the rout the Gipsey pointed out and found it a legible one to make a movement')[15] is hard to reconcile with the idea of madness. The slightly humorous tone of the opening soon gives way, however, to the impression of a desolate and disoriented man. The anxious resolve with which, after correcting an initial misdirection on the first day, he 'lay down with my head towards the north to show my self the steering point in the morning',[16] speaks of someone who is unsure of himself as well as of his route. Throughout his journey, Clare's difficulty in finding the way offers powerful evidence of how severely limited his accustomed geographical horizons were, but his mental distress plainly exacerbates the disorientation. Ronald Blythe is undoubtedly right in stating that in happier circumstances Clare 'would not have been either spiritually daunted or physically wounded' by the journey out of Essex.[17]

On the second day, Clare's growing tiredness is simply conveyed by the observation that 'I seemed to pass the Milestones very quick in the morning but towards night they seemed to be stretched further asunder'. Also, he is limping, 'for the gravel had got into my old shoes one of which had now nearly lost the sole'. He looks in vain for a straw bed, passing houses which, with darkness approaching, 'show the inside tennants lots very comfortable and my outside lot very uncomfortable'.[18] When one has read countless accounts by respectable tourists in this period of the ready hospitality that was routinely encountered in rural areas, nothing better expresses Clare's forlorn, outcast condition as a deracinated working man than this image of him gazing enviously at lighted cottage windows.

The next morning, he meets a gipsy woman who gives him advice on how he can shorten his journey using footpaths; but Clare, much of whose poetry is built from the daily experience of footpath walking, is under stress in foreign terrain and fearful of losing his way, and stays on the road. Further weakened by sleeping in a dyke, he takes a rest early the next day and hears a coach approaching; he writes at first that he 'cannot reccolect its ever passing', but later adds that 'The Coach did pass me as I sat under some trees by a high wall and the lumps lasshed in my face and wakened me up from a doze'. Clare's physical deterioration is increasingly apparent: by the time he gets to Stilton he is 'compleatly foot foundered and broken down', and he keeps on the move through Peterborough only from a sense of shame at sitting down. The identities of things, not to mention his own identity, get blurred: he eats grass by the roadside which 'seemed to taste something like bread'.[19] The journey out of Essex, in common with Clare's original departure from Helpston, is an example of the kind of involuntary travel which tends to 'muddle rather than define the persona of the traveller'.[20] It is the projected form of this self-alienation that comprises the well-known, tragic conclusion of Clare's account, when he fails to recognise his wife and children who intercept him in a cart on the outskirts of Northborough.

I have dwelt on this short prose text in order to point up the sharp differences between the experience of foot travel for people from contrasting economic backgrounds. This is walking, not pedestrianism. Although Clare's poetry, as we shall see, gives a wholly positive view of the rural walking which both structures and supplements his working week, the 'Journey out of Essex' is remarkable in providing, despite its idiosyncratic circumstances, a perspective on the bodily realities and mental effects of walking as it shaped the lives of the majority of the population. If space may rightly be perceived as genderised, then it is also socially stratified, and no analysis which fails to measure how class determines the nature and extent of human mobility within a given society can be theoretically secure.

In taking stock of a rural dweller like Clare, who rarely travelled far outside the village where he lived and worked, an important distinction arises between the mentalities of those whose knowledge is derived from constant perambulation of a small locality, and those whose socioeconomic position enables them to take a broader, more comparative view. John Barrell bases his pioneering study of Clare's poetry on just such an opposition between rural insiders like Clare, tied to a place 'which they had no time, no money, and no reason ever to leave', and the rural professional classes, who tended 'to think of a place almost completely in terms of its relations with other places', and whose literary spokesmen (such as Arthur Young) exhibit the generalising, objective approach characteristic of travellers through a region.[21] It is because of a certain 'postcolonial' guilt at the violence involved in latter-day varieties of this discourse that, James Clifford argues, ethnography has distanced itself from the traditional 'fieldwork' model of observation and experiment, which presupposes a 'primitive' organic culture visited by a modern urban researcher-explorer. He sides with the view that natives defined completely by the places to which they belong, 'unsullied by contact with a larger world', have probably never existed. This is part of the reason why he believes that 'travel', despite its evident class bias, is nevertheless the best term available for the study of cultures and their products under the general heading of displacement. Despite the manifest incongruity of speaking of John Clare as a traveller, I have taken the same approach in surveying the multiple forms of pedestrian passage, signalling, by the use of the umbrella term 'travel', that there are certain physiological and phenomenological features that unite Clare's local rambling with middle-class pedestrian tourism. Whilst not forgetting the fact that, as Clifford says, 'certain travellers are materially privileged, other oppressed',[22] this honours the commitment of the first generation of radical pedestrians, who regarded walking as a social equaliser.

DOROTHY WORDSWORTH: 'THAT WONDERFUL PROWESS'

It is well-known that Dorothy Wordsworth was a remarkable walker: this fact is implicit, for admirers and scholars of her work, in the quality of the observations of nature, of the diaristic fidelity to often minute and transient environmental stimuli, for which she is famous. It is taken for granted that these features of her writing stem from regular perambulation of an intimately known and partly domesticated terrain. However, little consideration has been given to her walking beyond this limited, if necessary, acknowledgement, and criticism usually passes speedily from the physical grounds of her creativity to the content of her perceptions. One important exception is the work of Meena Alexander.[23] Alexander, though, focuses chiefly on the local excursive walking recorded in the Alfoxden and Grasmere journals; in building on her intriguing speculations below, I aim to redress the balance and give at least equal attention to the travel writing.

In the domestic journals, walking is more a given of existence than a conscious aesthetic choice. This is not to say that Dorothy Wordsworth is unappreciative of the physical liberty she enjoys: in 1790, at Forncett, when still under the care of her uncle, she writes to Jane Pollard of her satisfaction that 'I have leisure to read; work; walk and do what I please', and in a later letter seeks to persuade her of the pleasures of twilight walking, though this appears restricted to the garden. At Windy Brow near Keswick in 1794, in the early days of domiciled life with her brother, she writes with renewed confidence of her 'wonderful prowess in the walking way', and responds sharply to criticism from her aunt that evidently reflected the prejudice against female pedestrianism: 'So far from considering this as a matter of condemnation, I rather thought it would have given my friends pleasure to hear that I had courage to make use of the strength with which nature has endowed me ...'.[24] It is unlikely that Dorothy Wordsworth, as a woman sensitive to the socially embedded character of the individual life, ever took these freedoms for granted.

The *Alfoxden Journal* of 1798, the earliest piece of continuous prose, albeit the most fragmentary, gives a powerful impression of an existence routinely structured around daily pedestrian excursions. Alexander and others have noted how many of the entries begin with the word 'walked'; typically, this introduces a paratactic sequence of observations, largely devoid of further verbal activity but dense with participial adjectives. They are raw, clustered notations, especially attentive to the changing appearances of sea and sky. At its most ascetic, her stripped-down writing accords to walking the status of an irreducible life-necessity, as in the following arresting series:

30th March. Walked I know not where.
31st. Walked.
1st April. Walked by moonlight.[25]

Part of the entry for 12 March, 'did not walk', was evidently unusual enough to warrant a mention.

In the *Grasmere Journal*, which was intended solely for the eyes of her brother, this instinct for mobility acquires a range of secondary aims. The overarching plot of the journal, as Susan Levin points out, is that of William's engagement and marriage to Mary Hutchinson,[26] and Dorothy's walking takes on new significances in the course of working through her complex feelings with respect to this major event. In May 1800, the month in which her journal begins – and begins, indeed, with William setting off to visit Mary in Yorkshire – , she records numerous walks to Ambleside to collect the post, anxious always to receive a letter from her brother (on the evening of his departure she exclaims, 'Oh! that

I had a letter from William!' [p. 16]). These walks, which more often than not end in disappointment, nonetheless appear to serve some psychic need: either they discharge in motion – desublimate – feelings that would become too oppressive in the confines of home, or they perform a symbolic reincorporation (a reaching-out-to or fetching) of the absent loved one into the home and community. Elsewhere, walking is tantamount to a love-service: in the early days of June, for example, her rambles are directed towards the digging-up of wild plants – globe-flowers, thyme, columbine, orchids, and others – and their transplantation to the garden of Dove Cottage, all with a view to the return of her brother, who then 'had an opportunity of seeing our improvements' (p. 25). At the other end of the journal, with the departure for Gallow Hill (and William's marriage) imminent, their walks carry undeniable overtones of wandering lovers:

> After tea we walked upon our own path for a long time. We talked sweetly together about the disposal of our riches [the forthcoming payment of the Lowther debt]. We lay upon the sloping Turf. Earth and sky were so lovely that they melted our very hearts. (p. 139)

In places like this one senses a need to affirm a unity of interest and experience in face of the approaching disturbance. It has been argued that Dorothy Wordsworth's imagination was not territorial in the manner of her brother's, that she did not take possession in writing of the places she visited or traversed – as he did, most famously, with Tintern and the Wye valley.[27] Her writing is territorial in another sense, to do with the high priority it accords to emplacement and personal security. In this context, walking may have the function of confirming boundaries in a non-possessive sense, of reaffirming and deepening a sense of belonging to the local environment in which a sheltered space has been discovered. The early poem, 'Grasmere – A Fragment', is a vivid illustration of this non-appropriative pedestrian practice. She begins the poem with an extended description of one particular cottage in the vale that excites her affections:

> when I sit on rock or hill,
> Down looking on the valley fair,

> That Cottage with its clustering trees
> Summons my heart; it settles there.[28]

The cottage she describes is no picturesque stereotype: its fields are 'rocky steep and bare' (l. 22), and only a thick screen of pines mitigates its exposure to the cold north wind; but it seems the habitation most in sympathy with its environment – or rather, as 'The very Mountains' child' (l. 20), it occupies the marginal position between natural freedom and domestic enclosure most consonant with the author's divided sensibility. The key word in the quotation above is, however, 'settles', which conveys the intense yearning for an ideal place of rest and security.

Reverting to that time in the past when she and William did, indeed, first come to 'settle' in the Vale of Grasmere, Dorothy then describes the first solitary walk she took as a stranger in the land:

> Lured by a little winding path,
> I quitted soon the public road,
> A smooth and tempting path it was,
> By sheep and shepherds trod.
>
> Eastward, toward the lofty hills,
> This pathway led me on
> Until I reached a stately Rock,
> With velvet moss o'ergrown.   (ll. 53–60)

It is under this rock, roughly textured and 'richly garlanded' with trees and other foliage, that she finds that winter has its own 'pleasure gardens' (ll. 62, 76); she therefore responds positively, as 'an Inmate of this vale' (l. 87), when a nearby streamlet seems to urge her to 'rejoice'. But the lines quoted above deserve further comment: they place the initiative firmly with the path, and it is an already-beaten path that is chosen (Dorothy Wordsworth's walks rarely take her on untrodden ways). Everything about her encounter with this new landscape appears to minimise her own intervention within it, and the jubilant conclusion is the voice of one who has found accommodation within the vale, without being so confident of her 'insatiable mind' (to quote William's parallel text, *Home at Grasmere*) as to find 'The Inmates not unworthy of their home'.[29]

'It is the property of the journal...to delineate passage', Susan Levin claims,[30] and perhaps the most commonly remarked-upon features of the domestic journals are those which give testimony to the discontinuities and overlapping demands of Dorothy Wordsworth's existence at Dove Cottage, her constant re-negotiation of the boundaries between household affairs, family life, community obligations and the activity of the literary circle of which she was herself a hesitant part-time member. In passages like the following, the entry for 12 November 1801, it is walking that comes closest to linking the disparate events of a typical day:

> A beautiful still sunshiny morning. We rose very late. I put the rag Boxes into order. We walked out while the Goose was roasting – we walked to the top of the Hill. M. and I followed Wm – he was walking upon the turf between John's Grove and the Lane. It was a most sweet noon. We did not go into John's Grove but we walked among the Rocks and there we sate. Mr Olliff passed Mary and me upon the Road – Wm still among the Rocks. The Lake beautiful from the Orchard. Wm and I walked out before tea – The Crescent moon – we sate in the slate quarry – I sate there a long time alone. Wm reached home before me – I found them at Tea. There were a thousand stars in the Sky. (p. 58)

In a series of short, elliptical sentences, this passage presents a quiet, restrained interplay between recorded motion (walking out together, William walking upon the turf, walking among the rocks, Mr Olliff passing, walking out after tea, William getting home first) and images of rest (the still morning, sitting among the rocks, the beautiful lake, the moon, sitting alone, tea-time, the stars in the sky). Here we have a good illustration of Anne Mellor's analysis of the distinctive rhetoric of the journals: a preponderance of metonymy, 'a destructuring principle occasioned by random contiguity, by happenstance', offset by figures of repetition. Such a rhetoric, Mellor suggests, constructs a non-unitary, relational self; 'an embodied consciousness, one that walks and sits, one that feels the warmth of the day'.[31] In the passage above walking is indeed the connective tissue that ties together the discrete actions, perceptions and feelings that make up an ordinary day, that *mobilises* experience into semi-coherence.

One of the highlights of the *Grasmere Journal* is Dorothy Wordsworth's description of her journey to Gallow Hill with her brother in July 1802. They travel by various means, including a post-chaise, but they also walk a good portion of the way. Close accounts of actual journeys such as this, evoking the serial experience of passage, are uncommon in literature of the period, so the narrative is memorable simply for encapsulating the everyday. The tiring uphill walk from Thirsk to Rievaulx Abbey, with the author increasingly footsore, as 'the Sun shone hot, [and] the little Scotch cattle panted and tossed fretfully about', is especially vivid. What is also interesting is the treatment the travellers receive at an inn in Thirsk: having arrived by post-chaise they are treated well, but 'when the Landlady understood that we were going to *walk* off and leave our luggage behind she threw out some saucy words in our hearing' (p. 148). This is a reminder of the prejudice and potential ill-treatment that awaited 'respectable' pedestrians on the road in England (less so in Wales and Scotland, it would seem) at this time; it is a rare moment in Dorothy Wordsworth's *Journals*, since she is not often removed from the integrative community of the Lakes, where she enjoys considerable freedom of movement, and where the strangers she encounters on her travels (as recorded in the *Journals*) are usually social inferiors. On her tours in Scotland and on the Continent, she is much less subject to harassment of this kind, deriving as it does from the peculiarly English need to codify all behaviour in terms of class. By the time of the first Continental tour in 1820, she is able to generalise that 'English Women (in quest of pleasure, at least) are certainly more adventurous than those of any other European Nation': on the way to Chamonix, she reports the case of a seventy-year-old woman 'who had walked over the whole of the steepest parts of the descent of the Col de Baume'.[32] This is one (frustratingly brief) piece of evidence of the popularisation of female pedestrianism abroad well before the better-known heyday of Victorian women travellers.

However, if more women were afoot in 1820, this transformation in travel practice was not so advanced that its benefits were undervalued, and a good deal of interest attaches in the Continental *Tour* to the author's moments of self-consciousness as a pedestrian traveller, her clear appreciation of its unique advantages and limitations. The party went by carriage for much of this tour (the lowland sections in France, Germany and the Netherlands), but almost the entire Alpine section was undertaken on foot, and this was for Dorothy the heart of the expedition: as they enter mountainous country and the path deteriorates, she is 'Happy in [her] freedom' (p. 135) when she and William are forced to dismount, having regarded the spells on horseback as a 'duty'. She is sensitive to the minor yet intense pleasures 'that belong only to the foot-traveller', and which punctuate a long day's walking with bodily respite: 'shady spots tempting us on, the predestined bourne of each small portion of labour, where, seated on the grass, or a mossy stone, refreshment came with rest' (p. 119). She values too the optimal freedom of movement that many other early pedestrian tourists comment on: descending to Airolo from the St Gotthard Pass, for example, the multiplication of footpaths tempt her to a 'vagrant course', oscillating between natural beauties and 'scattered huts' where human encounters take place (p. 192). She wavers too between the pleasures of being alone and the comfort of companionship: in a large enough group of walkers 'fresh society [is] always ready – and solitude to be taken at will' (p. 262), a flexibility that answers to her contrary impulses towards quiet self-possession, on the one hand, and relational affirmation, on the other.

There is a reverse elitism about pedestrian travel in many Romantic texts, and Dorothy Wordsworth is just as susceptible as her male contemporaries to comparative gestures which define the speaker as a *traveller* in contradistinction to those who, by virtue of their mode of transport, are mere *tourists*. Whilst still proceeding by charabanc from Interlaken to Lauterbrunnen, she envies the 'pedestrian liberty' that would prevent the 'sounding torrents' being disturbed by 'rattling wheels' (p. 116). Having claimed this liberty for herself, she is able to patronise less energetic and enterprising travellers – as when, ascending the Gondo Gorge, she adverts satirically to 'a carriage full of Gentlemen' who drove straight through a spectacular gallery, and who 'just looked aside at the torrent; but stopped not' (p. 259). It is during this final Alpine crossing, a backwards revisitation of William's memorable journey with Robert Jones on the walking tour of 1790, that Dorothy approves their having left the new Napoleonic road and thereby experienced 'the natural solitudes of the Alps unmastered by the equalizing contrivances of men' (p. 263). In this topsy-turvy rendering of the politics of walking, roads and road travel are seen as the quintessence of democratic republicanism, whilst pedestrianism appears the privilege of a self-satisfied minority who can practise their 'unmastered' identities in less cluttered surroundings. This perverse-seeming opposition is given some support by a number of incidents which underline the difference between a pedestrian tourist walking for pleasure and a local inhabitant for whom walking is part of a rugged and demanding way of life. Descending Mount St Salvador on a stony track, the group passes several women with unshod feet and 'burthens on their backs', provoking the comment: *'I* found it labour enough to steer my body *un*laden down the rugged way' (p. 212). Such observations offer a realist correction to the performative rhetoric of radical walking discussed in Chapter 2.

It is perhaps worth dwelling on one episode in the Continental tour, the ascent of the Rigi above Lake Lucerne, as a summation of many of the features discussed so far and a compelling, multi-layered narrative in which personal concerns are subtly filtered through the aesthetic experience of landscape. The account begins with a fine description of the author's pleasure in physical exertion and the visual shocks that a sudden break in the rhythm of walking can provide:

> *My* scheme was always the same, pushing right onward, till tired limbs made rest delightful; and I rose a hundred times with morning freshness. Mary, on the other hand, moved on perseveringly at a slower pace, with fewer intervals of repose. I often reclined against the hill, gathering bilberries, that hung, as plumbs do against a fruit-wall....Turning our faces from the mountain, and overlooking the bendings of the lake from Küssnacht to Lucerne, beyond the fair towers of that city, we beheld, though the air was hazy, an enchanting prospect....
> (p. 157)

The description of that prospect has something of a Claudean structure and tonality, but the realism of the walkers' varying pace, together with the sensuous detail of fruit-gathering, means that this picturesque 'station' is one that has been earned, and the enchanting serenity of 'low intermingled hills' and 'beds of hazy sunshine' seems aesthetic compensation for the labour of 'tired limbs'. Characteristically, however, Dorothy Wordsworth finds that that labour also brings rewards of a more social character: at the top of what was evidently a popular mountain, she enjoys the intrinsic companionability of high places, finding 'ladies, middle-aged men, students from the Universities...all met together as friends, as if being uplifted from the world did but bring human beings nearer to each other' (p. 158). In its inobtrusive way, this passage is a robust refusal of the conventional scenario of the sublime, which pits a lone individual against the forces of nature in a specular encounter that leads from self-diminishment through egotistic identification to self-empowerment. Here, by contrast, the contemplation of Alpine sublimities is a sociable experience – being 'uplifted from the world' has, paradoxically, a levelling effect.

The following morning, Dorothy and Mary Wordsworth return to the highest point of the mountain and admire a twilight scene of rapidly moving cloud and mist, alternately revealing and concealing the land below:

> The vapour, that had before travelled over different parts, now closed up all from our view in one instant: – nothing was to be seen but a sea of vapour. As rapidly, that vapour broke beneath our feet, hurrying on till it turned the hill of Rigi; and then dropped down to join a solid mass, condensed, and as white as snow, which was resting on the Lake of Zug. Again fresh volumes rushed on beneath us. It was a wonderful spectacle. Thick masses of clouds, and light smoky mists alike drove away with inconceivable rapidity, while, in the stiller air, the condensed vapour lay on the hills of Zug, as motionless as the snow on the far-distant mountains. What would not the exhibition-contrivers of London or Paris give for such power of rapid transformation!... Lake, houses, woods, fields, lost in the twinkling of an eye, – again appearing in full view, and yet again concealed! (p. 161)

This rapturous description expresses more than the obvious 'theatrical' interest of the interplay of transitory meteorological effects and huge, stable land-forms: it draws also on the fundamental contradictions in the writer's mental and emotional life already noted, on the powerfully ambivalent yearnings towards both mobility and emplacement that animate all her journal writing. Here these antithetical longings undergo a characteristic displacement onto the landscape, in a way that has not gone unnoticed by her critics.[33]

The theme of the impermanence of natural appearances in the passage above is premonitory of the incursion of death and decay into the narrative – preoccupations which, Susan Levin observes, are prominent in the Continental *Tour* generally.[34] Henry Crabb

Robinson, whom the Wordsworths met at Lucerne, had introduced them to two acquaintances, and all three thereupon joined the travelling group. It is on top of Rigi that the two young men take leave of them, bound for Zurich, and narrative chronology is interrupted at this point as the author notes that one of them, an American, was drowned three days later. This memorial provides a personal affective rationale for the following paragraph, where, recollecting the ruins of a catastrophic cliff-fall observed the previous day, she shudders at the thought of what would happen if the ground beneath her gave way: 'Below us, was a little chapel and a few houses: – *they* would be buried in the ruins – not to speak of *our* fate!' (p. 163). It is as though these thoughts of death can only be allowed entry into the text when they have some literal foundation in the time of writing, even if this means manipulating the order of events; Dorothy Wordsworth's fears may thereby evoke a common bondage to nature as destroyer, rather than a stereotypical terror in face of the sublime.

The sight of the green slope down which they must descend announces the conclusion of this episode. After contemplating the ruins of seemingly durable landforms (in the Rossberg avalanche), the sight of cattle with 'bells forever tinkling' (p. 163) immediately resuscitates in a gentler key the idea of eternity. The actual descent is eerily reminiscent of William Wordsworth's passage through the Gondo Gorge in *The Prelude*, Book VI, incorporating similar antitheses of movement and rest, constancy and mutability:

> In passing through the Rigi valley, perpetual marks of falling away or decay of mountains are visible, yet no fearful devastation; scars gradually wasting – and fragments tumbled down. Threads and ribbands of cataracts were now gently performing their work; but their paths on the declivities told a plain tale both of perpetual and fitful wasting. (p. 163)

Whereas her brother's reading of the 'Apocalypse' of nature as the 'workings of one mind' (VI. 567) causes it implicitly to redound to his own credit (to the capabilities of his imagination), Dorothy Wordsworth's experience has a humbler resolution, as she takes rest in 'a long wood-built shed, that looked towards an elevated painting of our Saviour bearing the Cross' (p. 164). However, she records the fact that whilst sitting in the shed she took out her journal, so in the shadow of her quiet acceptance of decay and death, and her contemplation of the Christian symbol of suffering and sacrifice, there lies a suggestion that the life of writing may compensate for the tragedies which befal the life of the body. In intimating her own physical dissolution in later years, her *Tour* is a traveller's promissory note to the 'prisoner on [her] pillowed couch'[35] she was soon to become.

There are three further aspects of Dorothy Wordsworth's walking and tour writing that deserve mention: her attraction towards certain classes of traveller whom she meets on the road; the role of association and memory in the plotting and performance of her pedestrian activity; and the opposition between wilderness and habitation as geographical co-ordinates of her contrary states of mind.

Linda Mills Woolsey has written of how Dorothy Wordsworth takes a keen interest in other women's lives to test Romantic concepts of personal freedom, and in particular of her fascination with itinerant women who were foreign to her own situation, but who symbolised a desired independence from a restrictive home environment. 'The *Grasmere Journals*, she writes, 'associate the precarious autonomy of the female vagrant with both the dispossession of the houseless woman and the freedom of the carman's lass, and identify the safe enclosure of domesticity with both the snug nest and the grave'.[36] Anita McCormick offers a more negative view of the narrative focus upon vagrant women, deciding that such characters 'are projections of Dorothy's worst anxieties: of losing her home, her role and her sense of self'.[37] I think the more equivocal interpretation is the more judicious one, since mixed feelings are plainly evident in most such encounters. Since Woolsey and McCormick concentrate on the domestic *Journals*, it is worth quoting a single passage from one of the expeditionary writings to witness the same ambivalence at work. On the first Scottish tour, when exploring Loch Etive and its shores, Dorothy and William meet a company of tinkers who use kilns, wherever they find them, as their sleeping quarters:

> The kilns are built of stone, covered in, and probably as good a shelter as the huts in which these Highland vagrants were born. They gather sticks or hether for their fire, and, as they are obstinate beggars (for the men said they would not be denied), they probably have plenty of food with little other trouble than

that of wandering in search of it, for their smutty faces and tinker equipage serve chiefly for a passport to a free and careless life.... They made a romantic appearance.... When we had landed at the other side we saw them, after having begged at the ferry-house, strike merrily through the fields, no doubt betaking themselves to their shelter for the night.[38]

These tinkers, shouting to each other in Gaelic, have already functioned as emblems of picturesque humanity, and here there is a measure of envious curiosity about their 'free and careless life'. On the other hand, the moral pressure behind the description of their 'smutty faces' and 'obstinate' begging is the signature of one who has made an emotional investment in domesticity and 'respectable' community service. Given the circumstances of Dorothy Wordsworth's life, these tensions are understandable, and Meena Alexander's specification of 'a knife-edge existence between changefulness that could grant freedom from entrapment and a displacement so radical that it could uproot and destroy'[39] is relevant and apt.

It is perhaps odd to suggest, when so much has been said about walking permitting self-renewal through fresh experience and the immediacy of encounter, that repetition and re-enactment are key constituents in most of Dorothy Wordsworth's pedestrian expeditions. But a deferral to past patterns of experience or their textual representation, or an insistent channelling of her personal journey of discovery along the beaten paths of precedent and secondhand associations, are marked features of her writing. The Scottish tour of 1803 was certainly a journey into new territory for Dorothy, William and Coleridge, though much of their route is plotted around what were by then well-established tourist destinations, so it is a re-enactment in cultural, if not personal, terms. (Without explicitly suggesting it, John Nabholtz provides strong evidence for Dorothy having supplemented her memory with details from Gilpin's Scottish tour.)[40] But the most prominent element of repetition or return in the *Recollections* is the persistence with which the author compares landscapes seen for the first time with the scenery of the Lakes: some copse-covered hills near Dumbarton put her in mind of Grasmere; Ben Lomond is compared unfavourably to Helvellyn; the rocky shoreline of Loch Ketterine reminds her of Lodore on Derwent Water; the view down Glen Coe suggests Patterdale and the head of Ullswater (pp. 239, 244, 272, 331).

Although such comparisons may be a natural human activity, it is difficult not to interpret overuse of them as part of the writer's characteristic reluctance to take imaginative possession of new territory: giving her home environment priority over the foreign is a mediated denial of her own prerogatives as a traveller.

On the Continental tour of 1820, Dorothy Wordsworth interprets the same tendency rather differently, saying that the sights of Italy give most pleasure when they evoke associations of her own country because 'Here we were but passengers of a day: *there* we must live and die' (*JTC*, p. 221). This devalues the fundamental experience of travel – the liberating surrender to a world of difference – and conveys unwillingness to set at hazard an identity deeply enwoven with her unique position in her family circle and community. A related aspect of this conservatism is the often exasperating reversion to William's walking tour of 1790 and its poetic record in Book VI of *The Prelude*. The dual, overlapping temporality gives Dorothy's narrative an honorific, recommemorative function that exaggerates her dependent status as sibling, traveller and writer. After seeing Milan, for example, the party returns to Lake Como purely because of the *Prelude* association, and finally discover the grail which had eluded them on their first visit: 'and there we came upon the track of the old road, the very *same* which my Brother had paced! for there was no other, nor the possibility of one' (pp. 245–6). The crossing of the Simplon Pass, which the travellers of 1820 execute in the reverse direction to Wordsworth and Jones in 1790, is an entire itinerary of mental traces. I do not wish to add much to Susan Levin's fine commentary on this episode, which analyses the complexities of re-enactment thoroughly and perhaps with more empathy than I can manage.[41] I do not detect the same strain of hostility in Dorothy's relations with William (though I accept that psychoanalytic categories are readily available to translate over-submissiveness into its opposite), and find the near-erasure of present experience by 'tales of thirty years gone by' (p. 259) a perfectly equable one in her narrative. I find Meena Alexander's idea that walking for William Wordsworth was a confident 'bodily signature' on the raw surface of nature, whereas his sister's repetition of that signature forced her 'to acknowledge her own secondary nature', a powerfully suggestive one. Alexander is writing with general reference to the Grasmere years, but her comments offer an apt summing-up of Dorothy Wordsworth's belated crossing of the Alps: she makes little effort to inscribe her own signature on a landscape rich with borrowed associations, but rather seems content to follow the beaten path of vicarious experience. If, as Alexander argues, 'To walk behind, to listen, would not discomfort the space she was allotted in the Romantic hierarchy',[42] the Continental *Tour* is a figurative, temporal manifestation of this genderised sense of personal space.

The binding influence exerted on Dorothy Wordsworth's tour writing by family loyalties and her heavy investment in domestic life is well demonstrated by the keen interest she takes throughout her travels in the forms of actual or potential habitation. In the *Recollections*, for example, the journey along the western shore of Loch Lomond yields the spectacle of a paradisal settlement – 'three or four thatched huts under the trees' lying opposite a large wooded island which gives the illusion of bounding a lake no bigger than Rydal Water, and encircling one smaller island. In fantasising a domestic idyll in this spot – one that ostensibly comprises only her brother and herself – Dorothy Wordsworth, with characteristic selflessness, dedicates the 'fairy island' as a retreat for William: 'he might row himself over with twenty strokes of the oars, escaping from the business of the house ...' (*RTS*, p. 246). One old larch, 'singled out for injury' (p. 247), however, constitutes the serpent in this paradise, a reminder that there is no absolute shelter in life and doubtless a metaphor of Dorothy's own 'human fears'. The highlight of the Loch Lomond visit, though, comes when they cross to the larger island, Inch-ta-vannach, and climb to its highest point. A 'sudden burst of prospect' shows the lake scattered with islands, with the whole scene 'in motion with travelling fields of light' (p. 251). There is a great variety of shape and surface among the islands; one has the appearance of a hermit's cell, another contains an 'ambiguous' ruin that could be mistaken for a tuft of trees, another is bordered by isolated trees that seem by an optical illusion to be vessels coasting the island. (Later, they land on one where there are several woodmen's huts, inspiring a primitivist fantasy of 'new settlers' [p. 254].) The climactic remarks on the hill-top prospect embody a rare, memorable reflection on the mental effects of landscape:

> a new world in its great permanent outline and composition, and changing at every moment in every part of it by the effect of sun and wind, and mist and shower and cloud, and the blending lights and deep shades which took the place of each other,

traversing the lake in every direction. The whole was indeed a strange mixture of soothing and restless images, of images inviting to rest, and others hurrying the fancy away into an activity still more pleasing than repose; yet, intricate and homeless, that is, without lasting abiding-place for the mind, as the prospect was, there was no perplexity.... (p. 253)

The opposites that are beautifully harmonised in this passage – permanence and change, movement and rest, homelessness and abidance – are those which, I have argued, act as a structuring principle of Dorothy Wordsworth's travel writing, and which articulate her essential ambivalence towards travel itself. 'Pedestrian liberty', which she practised and enjoyed to an extent not available to many women of her period, provided release from domestic responsibilities and narrow social definitions of her role, and mobilised her talents as a writer; but, for reasons of personal insecurity and cultural disempowerment, she could not embrace an individualising, self-aggrandising liberty as unreservedly as her brother, and so she habitually reaches out, in text as well as life, to the fragile promise of some 'lasting abiding-place'.

JOHN CLARE'S PERAMBULATORY POETICS

I have already pointed to the apparent anomaly of categorising John Clare as a pedestrian 'traveller'. Indeed, Alan Bewell, reviewing Anne Wallace's *Walking, Literature, and English Culture*, objects to her inclusion of Clare on the precise grounds that he was 'a great walker' but 'hardly a traveller'.[43] I took the view that this was a legitimate and worthwhile step, in that it recognised certain inalienable features of pedestrian practice whilst still allowing for a differentiation of types of walking. Clare himself, of course, is acutely aware of the varying character and social distinctions of different forms of locomotion. 'Out of Door Pleasures' is one of many poems which catalogues with great affection the mixture of natural perceptions and ordinary human incident available to the footpath-walker; by contrast, 'A huge cloud of dust all the coaches conceals / They are hid in the smoak that flies up from the wheels'.[44] The cloud of dust implicitly bars the passengers' visual access to the rural pleasures that Clare celebrates. The essential foreignness to Clare of such forms of passage, and their destabilising mental effects, are suggested by the section in his 'Autobiography' describing his first trip to London:

> by seeing people at my old occupations of ploughboy and ditching in the fields by the road side while I was lolling in a coach the novelty created such strange feelings that I coud almost fancy that my identity as well as my occupations had changd that I was not the same John Clare[45]

However, Clare's poetry does not support a clear terminological distinction between walking and travel. A poem like the superb late sonnet, 'I Am', it is true, marshals such an opposition in contrasting the earth-bound prison of experience with the unshackled soul of his earlier life: his body now 'plod[s] upon the earth', whereas he once possessed 'A spirit that could travel o'er the space / Of earth and heaven' (ll. 2, 9–10). 'Travel' here connotes freedom, privilege, self-confidence. Yet 'Evening', from the same period, uses the same word with respect to practical journeying:

> It is the silent hour when they who roam,
> Seek shelter, on the earth, or ocean's breast;
> It is the hour when travel finds a home,
> On deserts, or within the cot to rest.   (ll. 1–4)

Perhaps some shadow of the atrophied root word, 'travail', is present in Clare's usage, which would strengthen the case against linguistic oversensitivity on this issue.

Nevertheless, it is proper to take stock of the particular significances which walking held for Clare, and of the place it had in the rural communities to which he belonged. Ronald Blythe, in stressing the indivisibility of Clare's poetry from the regular practice of footpath walking, very helpfully points to the ambiguities of this practice in context, appearing as both normal and exceptional, routine and eccentric:

> Just before this century, everyone walked. Clare's constant walking in his landscape was the norm; except that sometimes he walked, where his Helpston neighbours were concerned, to what was recognisably work – gardening, ploughing, hedging, erranding; and sometimes to what to them was clearly not work – reading and writing, in dips and hollows – a very strange thing

to do; and sometimes he walked just to look. And so he became what most village people dread being: odd, strange, different.[46]

Given the higher density of population in the countryside in the early nineteenth century, footpaths, Blythe notes, 'did not guarantee solitude', so Clare's eccentricities were difficult to conceal from his fellow-villagers. The 'Autobiography' relates how he had more opportunity as a field-labourer than a gardener to 'set down my thoughts', and how he frequently missed church on Sundays because he would then have more chance of being alone in the woods or on the heaths. Since he was not generally known as a poet there was plenty of speculation about his 'odd' behaviour: 'they fancied I kept aloof from company for some sort of study others believd me crazd and some put more criminal interpretations to my rambles and said I was night walking assosiate with the gipseys robbing the woods of the hares and pheasants'.[47] Clare, in fact, spent a good deal of time with gipsies and, though he was critical of aspects of their morality, sympathised with the prejudice and discrimination they encountered and thought them less dangerous than 'some in civilizd life'; indeed, he claims that he 'became so initiated in their ways and habits that I was often tempted to join them'.[48] As James McKusick has argued, the asylum years deepened in Clare a lifelong sense of class solidarity with gipsies and other vagrants, an identification that reflects his perception of himself as an outsider and the inhibitions of his territorial way of life.[49]

It would therefore be simplistic to contrast Clare with other Romantic pedestrians studied in this book on the grounds that he belonged to a rural class for which walking was an integral part of the communal way of life. Questions of where, when, and to what purpose one walked were crucial in determining how this activity was regarded; that Clare's rambling could be viewed as 'odd', 'crazy', or even 'criminal' by his neighbours proves the fallacy of quick generalisations about class experience and attitudes. In a rural community like Helpston, walking was the regular way to move about the parish, to go to and from work, to go to church, and so on; footpaths were also used recreationally, and for lovers' walks – Clare himself recalls such love-rambles with his wife-to-be, Patty, in his 'Autobiography'.[50] Bob Bushaway writes of another function that walking performed in rural communities in this period – the ritual marking of parish boundaries in Rogation week. This custom, known as 'beating the bounds' or 'perambulating the parish', had both the secular purpose, in a pre-enclosed landscape, of guarding against encroachments and preventing the destruction or obscuring of field boundaries, and the religious aim of fostering self-restraint and mutual respect:

> During the perambulation, the community was enjoined to 'consider the old ancient bounds and limits' so that each individual should 'be content with our own, and not contentiously strive for others, to the breach of charity, by any encroaching one upon another, or claiming one of the other further than that in ancient right and custom our forefathers have peaceably laid out unto us for our commodity and comfort.'[51]

The only allusion to parish perambulation in Clare's work that I know of comes in the 'Letter to The Every-day Book', which mentions the practice of digging holes along the boundary, which were filled with stones or, more oddly, children:

> On Holy thursday they go round the fields opening the [meres] or land marks w[h]ere they still keep up an ancient custom [of] scrambling in the mere holes for sugar plumbs & running races for cross skittles in which old & young often join... young boys & girls the sons & daughters go on purpose to be placed on their heads in the mere[52]

A side-effect of the perambulations was to provide the community with 'a mental map of the parish', which became its 'collective memory', and this invites comparison with Clare's poetic record of daily perambulation of field and heath in the pre-enclosed landscape of Helpston, and to its evocation of the 'circular' sense of place which, according to John Barrell, such a topography encouraged in those who knew it intimately as insiders.[53]

Perhaps the most important thing to remember from the perspective of class is that, for Clare, walking, or the kind of walking he valued most, signified freedom from labour. This is conveyed with eloquent directness in the early poem, 'Sunday Walks', which contrasts the working week with the tranquillity of the sabbath, when everything in nature acquires 'brighter colors' (l. 21). For the speaker of the poem, non-utilitarian walking carries a surplus value, and this value is transferred as aesthetic credit to the environment. As though the weekly cycle is marked by a self-conscious shifting of subject positions, the 'six days prisoner' (l. 59) steps outside his practical workaday mentality and becomes a natural religionist:

> He ponders round and muses with a smile
> On thriving produce of his earlier toil
> What once was curnels from his hopper sown
> Now browning wheat ears and oat bunches grown
> And pea pods swelld by blossoms long forsook
> And nearly ready for the scythe and hook
> He pores wi wonder on the mighty change
> Which suns and showers perform and thinks it strange
> (ll. 41–8)

The respect for Nature's God ('the power...Who rules the year' [ll. 53–4]) is thrown into relief by the perfunctory reference to the priest doing 'whats requird' and the glimpse of the 'godly farmer' shutting his Bible by evening to 'talk of profits from advancing grain' (ll. 83–6). The formal language of worship is then contrasted with the higher prayerfulness of the ordinary self-voicing of nature:

> Then leave me sundays remnant to employ
> In seeking sweets of solitary joy
> And lessons learning of a simpler tongue
> Were nature preaches in a crickets song
> Were every tiney thing that lives and creeps
> Some feeble language owns its prayer to raise
> Were all that lives by noise or silence keeps
> An homly sabbath in its makers praise
> There free from labour let my musings stray
> Were foot paths ramble from the public way (ll. 91–100)

Despite the obvious risks of pious sentimentality or mock-heroic whimsy, this is not a conventional neoclassical rhetoric of animated nature, but a genuine attempt at a non-anthropocentric vision of the holiness of life: the speaker is eavesdropping on the 'prayers' of creatures who constitute their own legitimate centres of being. The poet's ability to listen in, his humility in face of the 'homly sabbath', is conditional upon the undetermined, hence pleasurable, character of his walk. And if freedom from labour is the prerequisite for his environmental awareness, it is also the ground of his creativity: it is in these wanderings off the 'public way' (l. 100), following the 'winding baulks' (l. 103) or roaming at will over the 'trackless' hills and heaths, that he acquires the facility to 'add a song' to those of 'natures minstrels' (ll. 137–8) – thereby claiming, even while diminishing, his share in the aesthetic produce of leisure.

The 'baulks' (grassy paths) of which Clare speaks were typically lost as a result of enclosure, as the open-field landscape in which he grew up, and which provides the nostalgic terrain of many of his mature poems, was ploughed up and parcelled out in larger and more economically manageable units. There is now a good deal of fine criticism on Clare and enclosure, to which I have nothing new to add; but it is necessary to register its importance in an assessment of Clare's relation to the development of peripatetic. John Goodridge and Kelsey Thornton have assisted significantly in this task in an essay exploring the concept of trespass in Clare's work. Illuminating their account with reference to the brutal provisions of the Trespass and Game Laws (that were actively enforced in the period), and to Clare's realistic fears of persecution by the rural authorities, the authors throw light on a wide range of literal and figurative boundary-breakers in his poetry – from ants to tramps, from moles to gipsies. Given the danger of identifying openly with acts of trespass, the figurative avenues – more mediated, but less censorable – are particularly important, as are symbolic objects (like broken gates) that can fuel 'fantasies of penetration and escape'. What is established is a passionate and long-lasting, though often subtly encoded, interest on Clare's part in the politics of pedestrianism: as Goodridge and Thornton observe, 'We think of him always walking trackless ways; yet we have to realise how much that was a deliberate, often difficult and dangerous thing to do.'[54]

A few strong tokens of this interest will suffice. In 'The Lamentations of Round-Oak Waters', the stream's lament for the despoliation of the land by improving landlords includes mention of the baulks which crisscrossed the pre-enclosed land; the loss of these public paths is seen here from a practical rather than recreational point of view:

> The gentley curving darksom bawks
>   That stript the Cornfields o'er
> And prov'd the Shepherds daily walks

> Now prove his walks no more
> The plough has had them under hand
> And over turnd 'em all
> And now along the elting Land
> Poor swains are forc'd to maul   (ll. 101–8)

'Lamentations' is an elegy, but political persuasion is a traditional constituent of the elegy, and by the end of the poem its tone of complaint has given way to a hectoring rhetoric of accusation directed at the ruling-class originators (rather than their agents) of the rapacity deplored: the 'aching hands' that axed the riverside willows were not those 'that own'd the field / Nor plan'd its overthrow' (ll. 169–72). 'The Fens' concludes with a similar protest against agrarian 'gain' and environmental degradation, but in this poem the introduction of a political theme definitely unsettles its generic relations: it begins as a gentle rural walk-poem, with the speaker 'Wandering by the rivers edge' (l. 1) and observing diving kingfishers, dabbling ducks and 'gossiping' geese; it is only in the final third, as he takes a wider view over the 'naked levels' (l. 66), that an animus develops against the farmers responsible for creating an unnaturally flat, featureless and monochrome landscape:

> And muse and marvel where we may
> Gain mars the landscape every day
> The meadow grass turned up and copt
> The trees to stumpy dotterels lopt   (ll. 83–6)

The poet's absorption with this theme has the effect of halting the text's progress, as well as his own represented passage over the land: 'Change cheats the landscape' (l. 92), and cheats the reader's initial expectations of the poem.

It is in 'The Mores', arguably Clare's finest poem, that the politicisation of walking in the context of enclosure achieves its most powerful and concentrated expression. In the first two thirds of this poem Clare's recollections of the pre-enclosed landscape are described from a stationary viewpoint that takes in the whole of an immense 'level scene' (l. 1), and the passages of personal nostalgia or political invective do not interfere with the sweep of this fixed, circling gaze. In the final third, however, he approaches his subject from the perspective of someone moving through that scene, projecting the difficulties posed by the landscape of old to the pedestrian traveller as a source of fortunate error:

> Each little path that led its pleasant way
> As sweet as morning leading night astray
> Where little flowers bloomed round a varied host
> That travel felt delighted to be lost
> Nor grudged the steps that he had taen as vain
> When right roads traced his journeys end again
> Nay on a broken tree hed sit awhile
> To see the mores and fields and meadows smile
> Sometimes with cowslaps smothered – then all white
> With daiseys – then the summers splendid sight
> Of corn fields crimson oer the 'headach' bloomd
> Like splendid armys for the battle plumed
> He gazed upon them with wild fancys eye
> As fallen landscapes from an evening sky
> These paths are stopt – the rude philistines thrall
> Is laid upon them and destroyed them all   (ll. 51–66)

I have deliberately quoted at length in order to reproduce faithfully the shuddering halt imposed on the reader by the triple-plosive energy of 'stopt' in line 65: after the loose, expansive syntax of the preceding lines describing the compensations of going astray, the legal violence of being dispossessed of traditional rights of way is vividly simulated by that sudden interruption of rhythm and sense. The particular experience of landscape which is thus peremptorily consigned to history has been well reconstructed by John Barrell:

> The system of roads in an open-field parish... is in the first place for the circulation of men and cattle within the parish.... [T]here will often be no particular distinction between a road that will in fact lead the traveller, however indirectly, out of the parish, and one which leads the farmer to his land and the labourer to his work but takes the traveller nowhere. For [an outsider] the road-system of an open-field village is a labyrinth, whose secret cannot be learned without a guide....[55]

Clare's delighted traveller found Helpston such a labyrinth; Clare, as an insider, held the key to it. But the result of enclosure is to unravel the labyrinth, replace it with a restricted number of thoroughfares in the spirit of the new regional and national communications network, and erect physical and legal obstacles to passage outside what is now sharply delimited public space: pedestrian choices for any given journey before enclosure were typically plentiful, but enclosure drastically curtailed and channelled the mobility of villager and traveller alike, as countless 'paths to freedom and to childhood dear' were marked with signs saying ' "no road here" ' (ll. 69–70). The legacy of this policy of legal expropriation has continued up to the present day in Britain, with campaigns for increased access to privately owned land and periodical acts of mass trespass; Clare's bitter denunciation of the consequences of 'lawless laws enclosure' (l. 78) for physical access to, as much as the appearance of, the countryside, makes 'The Mores' a symbolic act of trespass which forcefully anticipates that tradition of civil dissent.

Given the unique centrality of local rambling and footpath-walking to the subject-matter of Clare's work, it is to be expected that the pedestrian origins of his verse should have consequences for its aesthetic form as well. The most obvious such consequence is a predisposition, in all but his shortest poems, for open-ended seriality in description, for what Seamus Heaney has admired as his 'love for the inexorable one-thing-after-anotherness of the world'.[56] I have already discussed this propensity for progressional ordering of the objects of perception as a characteristic of peripatetic form in Chapter 3. Clare manifests it in an extreme form, and, unsurprisingly, it has not always found favour with his critics, then or now. John Taylor, his first publisher, advised him that he should be more selective in his descriptions and treat the appearances of nature more 'philosophically'; he complains that one of his longer poems 'rambles too much' – an almost dead metaphor which nevertheless demonstrates the critical prejudice which walking poems encounter. The very principle of their structure is a sign of deficiency in the eyes of conventional poetics. Very much more recently, Roger Gilbert has contended that good walking poems must transcend mere journalistic fidelity to observation in order to conform to the different ontology of art; Clare's 'repeated insistence on the value of the experiences he records fails to confer an equivalent value on the poems'. On the other side of the debate, John Barrell values Clare's 'aesthetic of disorder' as a counterblast to the rigidities of eighteenth-century landscape aesthetics: substituting multiplicity and particularity for Thomsonian order and distance, Clare praised landscape 'on account of its formlessness, its failure to accommodate itself to correct taste'. Somewhat modifying this harsh opposition, Timothy Brownlow has compared Clare's poetic perambulations to the perimeter walk in eighteenth-century landscape gardens, which exploited shifting viewpoints and the blurring of garden and country permitted by the ha-ha. He notes, though, that his walks are 'random and informal in a way that the eighteenth-century poetry of perambulation seldom is'.[57]

A good example of the serial structure of Clare's walking poems is 'Reccolections after a Ramble'. Over some 250 lines, we follow Clare along a path by the side of a river, observing a descending lark, bees heavy with cowslip pollen, and 'Brazen magpies full of clack' (l. 17); over a bridge, from where he sees young men bathing naked, embarrassing a passing girl, and hears singing from a gipsy camp; through a wood, where he lies down to rest, breaking an oak bough to fan flies from his face; up a hill, pausing now and then to count distant church spires and to satisfy his 'curious eye' (l. 102); into a blackthorn bower to take shelter from a shower, relishing meanwhile the effects of the changed weather on the plant and animal life around him; and across the fields on his return journey, seeing all around him the 'freshning plains' revived by rain just as 'labour' (a recurrent metonym in this poem) is revived by ale (ll. 157–60). The poem ends with an assertion of the 'wisdom' to be gathered in such rambles, and elevates his rural muse above 'epics war harps' (l. 256).

These final statements, however, do no more than gesture at a teleology of the walk or poem. They are too perfunctory to perform that passage from the sensuous to the intellectual, from concrete to abstract, from the outer to the 'inward eye', that is typically celebrated in Wordsworth. The whole substance of the poem, and its residing impression in the mind of the reader, is of a loose train of naturalist's observations, of which the following lines on the aftereffects of rain are representative:

> And upon the dripping ground
> As the shower had ceasd again
> As the eye was wandering round
> Trifling troubles causd a pain

> Overtaken in the shower
> Bumble bees I wanderd bye
> Clinging to the drowking flower
> Left without the power to flye
>
> And full often drowning wet
> Scampering beetles racd away
> Safer shelter glad to get
> Drownded out from whence they lay
> While the moth for nights reprief
> Waited safe and snug withall
> Neath the plantains bowery leaf
> Where there neer a drop coud fall   (ll. 137–52)

Although Clare's images are not without traces of sentimental anthropomorphism, the creatures he describes have their separate and legitimate existence, and are not swept into service in some overarching plot of the poet's consciousness. The place that the various observations and experiences Clare assigns to his ramble have is the place which they are assumed to have occupied on the line of the walk. No doubt this lack of design is artefactual in its own way – a rhetoric of informality and contingency may be no easier to achieve than a rhetoric of unity-in-complexity – but it is a designless design which aptly renders the effect of an open, unpremeditated movement through space, permeable to experience and connecting with the world passionately yet unobtrusively.

Seriality and multiplicity may be the keynotes of Clare's perambulatory poetics, but they are modified in various ways in a good number of his poems. One such modification is suggested by Barrell's intriguing characterisation of his 'unimproved, open-field imagination'. For Barrell, Clare's best poems – 'Emmonsails Heath in Winter' is his chief example; I would put forward 'Out of Door Pleasures' as another – embody a 'circular' sense of space unique to the pre-enclosed landscape in counties like Northamptonshire – a sense 'of being able to see the whole field, and the whole of what was being done in it, at one *coup d'oeil'*. That is, because vistas are uninterrupted by hedges or fences, and because the densely local quality of his knowledge means that distance – whether of natural objects or human activities – does not imply lesser particularity, Clare's loosely conglomerative syntax can strike the reader not so much as a 'continuum of successive impressions' as a 'complex manifold of simultaneous impressions'.[58]

This is a powerful rationalisation of the paratactic structure of Clare's verse, and a useful rebuttal of criticisms of his indiscipline and self-indulgence, but it is unlikely that it can convincingly be applied to more than a small number of poems. Another form of modified seriality consists in a progressive narrowing-down of perspective: 'The Robin's Nest', for example, is a poem which relieves the undermotivation of rural walking with a quest motif, as the eye of the text slowly moves from the wider evidence of spring and the poet's happiness at leaving behind 'the ruder worlds inglorious din' (l. 6), to the woodland home of the robin and the protection of 'old neglect' (l. 50), to its perching position on the 'dead teazle burs' (l. 71) and, finally, to its nest on the ground by a mossy tree stump. A third type of concession to the principle of order is the use of a focus incident, which helps to differentiate a walk in memory and within the canon of the poet's works. A minor example would be the sonnet on 'The Squirrel's Nest', in which the speaker's 'pleasant walk' is enlivened by the sight of an oddly-shaped nest which he assumes must belong to some 'foreign bird'; on climbing the tree to satisfy his curiosity,

> somthing bolted out I turned to see
> And a brown squirrel puttered up the tree
> Twas lined with moss and leaves compact and strong
> I sluthered down and wondering went along   (ll. 11–14)

Unspectacular as this *dénouement* is, it nevertheless provides the poem with the rudiments of a plot: it is a minor revelation that passes for a destination, and prepares for closure. However, the techniques I have described – circular space, narrowing-down of perspective, and focus incidents – vary, without fundamentally correcting, Clare's dominant aesthetic, which, in its commitment to seriality, multiplicity and particularity, affects to disown the colonial gaze of the Romantic nature-poet in favour of the 'simpler tongue' of things-as-they-are-experienced.

The emphasis on the neighbourliness of familiar objects in the rural environment, noticeable in several of the quotations from Clare's work above, raises the question of Clare's relation to the current debate around 'Romantic ecology'. In British criticism, the keystone of this debate is Jonathan Bate's slim 1991 book, *Romantic*

Ecology, which makes intermittent reference to Clare; Bate resumes the argument, with a single critical focus, in an essay published in a special issue of the *John Clare Society Journal* in 1995.[59] While I cannot attempt here to engage the issue in a sustained way, it is worth briefly acknowledging its relevance to my discussion. It is widely felt, as I noted in Chapter 3, that walking is the most environmentally-sensitive mode of travel, as it restores a sense of proportion between traveller and environment obscured by other forms of transport, and because it minimises the traces of its passage. In the late twentieth century, this claim carries a political charge which it would be unwise to try to read back into the life and culture of Clare's period. In the years following enclosure the parish perambulations I described earlier sometimes turned into acts of civil disobedience – Bushaway comments on an incident at Otmoor in 1830, when a thousand people repossessed the moor, tearing down 'fences, hedges and every obstacle in their path'.[60] While I was writing this chapter, in summer 1996, green activists effected the closure of the M41 in London, staging a massive street party and planting trees in the fast lane, in a protest that was in part an assertion of the pedestrian rights of city-dwellers against the hegemony of car culture. In both historical situations one can observe how urgently walking has been politicised, but as much is lost as gained in blurring the differences between the two. If the political significance of walking, for Clare, hinged on questions of land ownership and rights of access, it now hinges on much larger questions of environmental degradation, sustainable ways of life, and the rejection of industrial-capitalist society. By the same token, just as it would be absurd to suggest that modern Londoners are making a political statement every time they set foot outside their doors, so it needs saying that much of Clare's local rambling around Helpston and Northborough had no political meaning or intent.

Having stated the dangers of dubbing the literature of the past with today's political script, it is still possible that the literature-ecology debate may illuminate aspects of Clare's treatment of human interactions with the natural environment. A key concept that can help test this utility is that of *place*. The 'sense of place' was, as I have recalled, the principal subject of Barrell's study of Clare's poetic syntax; but Barrell's construction of Clare's place-sense, however much it is an intimate, participatory, insider's mentality, would receive unmerciful scrutiny from the more ardent apostles of ecocriticism. Lawrence Buell's magnificently learned and thought-provoking book, *The Environmental Imagination*, articulates the problem well. For Buell, one of the criteria of an ecological text is that the non-human environment should constitute an independent source of interest and value, and should not figure solely as a setting for human action or a symbolic resource for anthropocentric desire. A commitment to place, he argues, does not guarantee ecological soundness, since places may acquire powerful meanings embedded in human interests and values, which may in turn insulate the observer from awareness of their intrinsic qualities. At times Buell's prosecution of this argument borders on the perverse, as when he castigates Hardy's *The Return of the Native* for attending too little to the niceties of the Wessex ecosystem; but elsewhere he achieves a more balanced view which acknowledges the necessary reciprocities of subject and object:

> All creatures process their environment subjectively and seek to modify it in the process of adapting to it. It is not a question of whether we can evade this ground condition but of how to make it subserve mutuality rather than proprietary self-centeredness.[61]

As part of this cultural programme, it is interesting that he takes time to reclaim personification from critical distaste, on the grounds that it performs a blurring of hierarchical boundaries between human and nonhuman more in tune with twentieth-century science than with nineteenth-century aesthetics. However questionable its uses in such ecological phenomena as James Lovelock's Gaia hypothesis or the mother-earth mythopoeia of feminist theologians, 'the image of nature's personhood', according to Buell, has the laudable potential to mobilise an 'ethics of care'.[62]

In the present context, these remarks cannot help but bring to mind the lines in Clare's 'The Flitting', the poem that records his feelings of sadness and loss at moving the small distance from Helpston to Northborough, which speak of 'Strange scenes' being 'Vague unpersonifying things' (ll. 89–90). What Clare laments is a sense of belonging and deep interconnectedness that arose within a very circumscribed rural existence, in which the most ordinary and ephemeral appearances could excite the domestic affections: 'every weed and blossom too / Was looking upward in my face / With friendships welcome "how do ye do"' (ll. 126–8). However sentimental this may seem to modern readers, it is a genuine expression of a nature sensibility for which the appropriate metaphors are indeed those of companionship and neighbourly love – which does not read, that is, as subjective imposition on the environment. It is the task of Clare's perambulatory muse, 'Who walks nor skips the pasture brook / In scorn' (ll. 165–6), to memorialise this sense of community with the nonhuman, since, though green things everywhere have a consolatory power, the outcome of his relocation is to dilute the personhood of nature:

> The summer like a stranger comes
> I pause and hardly know her face (ll. 3–4)

The intense lococentrism negatively evoked through the experience of geographical displacement, and which was also disturbed, as we have seen, by the disfiguring effects of enclosure on the landscape, is eloquently rendered in Clare's own account of his first exploration of Emmonsales Heath as a child, expecting to find the end of the world on the horizon:

> So I eagerly wanderd on and rambled among the furze the whole day till I got out of my knowledge when the very wild flowers and birds seemd to forget me and I imagind they were the inhabitants of new countrys the very sun seemd to be a new one and shining in a different quarter of the sky still I felt no fear my wonder-seeking happiness had no room for it I was finding new wonders every minute and was walking in a new world[63]

His experience of this 'new world' has the effect of defamiliarising the 'old' world of the village on his return, though it is implicit in the notion of intersubjective relations with plants and animals that he would relearn his kinship with the natural 'inhabitants' of Helpston. His early poem on 'Emmonsales Heath' shows that the new world was also, eventually, assimilated to his insider's mental map of the district. There is repeated emphasis in the first half of this poem on the fact that the heath has so far happily withstood the march of agrarian reform: it 'still' lingers in its 'wild garb' (ll. 1–2); 'Stern industry... Still leaves untouched [its] maiden soil' (ll. 9–12); the furze, heather, bracken and grass are the same as they have always been (ll. 29–32); and a brook 'Still wildly threads its lawless bounds' (l. 35). The idea of nature's self-determination is reinforced by images of non-human domesticity: birds find shade to build their nests, and flowers 'Find peaceful homes' (l. 20), on land neglected by man. This representation of nature as a 'family' (l. 17), which shares its advantages with man but otherwise keeps itself to itself, is – if one accepts that there is more to such language than stale sentimental conceits – as 'ecological' a vision as Clare provides in his poetry. The fact that Clare can reproduce as an adult his 'boyish walks' (l. 65) on the heath means that he can re-experience this commonality in the only place he felt truly at home in the world. It is also a way of rehearsing the territorial limits of his imagination, a personal beating of the bounds which, elsewhere and at other times, he could re-enact only through memory and the perambulations of verse.

7
Walking and Talking: Late-Romantic Voices

> I cannot see the wit of walking and talking at the same time.
> William Hazlitt

Thus Hazlitt asserts at the beginning of one of the greatest short pieces of walking literature, his essay 'On Going a Journey', from *Table Talk* (1821). Barring some inimitable digressions, the opposition of solitary and accompanied walking is the running theme of the whole essay. To begin with, the choice of whether 'to talk or be silent, to walk or sit still, to be sociable or solitary'[1] is emphatically one-sided: for Hazlitt, 'The soul of a [pedestrian] journey is liberty, perfect liberty' – not just the bodily, navigational freedom celebrated by many early pedestrian tourists, but the corollary liberty 'to think, feel, do just as one pleases' (p. 136); and this can be embraced only, it seems, when one is alone. Because it is so subjectively and privatively conceived, such liberty distrusts the interpersonal character of language praxis, disdains the 'awkward silence, broken by attempts at wit or dull commonplaces' (p. 137) that marks language reduced to the merely phatic. Indeed, that realm of automated verbal behaviour, which threatens to 'speak' the subject, is precisely what one is escaping from: in motion, 'We are no more those hackneyed common-places that we appear in the world' (p. 142), but instead, presumably, to extend the metaphor, nonce-words, or figurative deviations from our literal selves.

Hazlitt is pessimistic regarding the possibilities of social sympathy on the road, and believes that any communicative effort jeopardises his temporary self-fashioning as a 'citizen of the world' (p. 141) and reels him back to the containing facts of personal history. As a pedestrian traveller, 'it is great to shake off the trammels of the world and of public opinion – to lose our importunate, tormenting, everlasting personal identity in the elements of nature, and become the creature of the moment, clear of all ties' (pp. 141–2). For Hazlitt, walking unhinges the socially constructed and maintained self, places it in suspension, allowing the mind to become a screen on which the passing image is momentarily projected, overdubbed with ideas and memories generated according to associationist principles:

> We cannot enlarge our conceptions; we only shift our point of view. The landscape bares its bosom to the enraptured eye; we take our fill of it; and seem as if we could form no other image of beauty or grandeur. We pass on, and think no more of it: the horizon that shuts it from our sight also blots it from our memory like a dream.... So in coming to a place where we have formerly lived and with which we have intimate associations, every one must have found that the feeling grows more vivid the nearer we approach the spot, from the mere anticipation of the actual impression: we remember circumstances, feelings, persons, faces, names, that we had not thought of for years; but for the time all the rest of the world is forgotten! (pp. 144–5)

Hazlitt's slighting of memory here perhaps has the effect of disguising the Wordsworthian tenor of his reflections: his projected return to Llangollen, which he had first visited as an ardent Republican in 1798, is clearly modelled on Wordsworth's return to the Wye in 'Tintern Abbey', and this filiation is confirmed by actual allusions ('Yet will I turn to thee in thought, O sylvan Dee', 'the picture of the mind revives again'). Superficially, at least, his representation of a mobile consciousness is more positive than Wordsworth's, because it figures movement spatially rather than temporally: change of place tends naturally to be perceived as gain, passage of time as loss. 'In setting out on a party of pleasure', he states, 'the first consideration always is where we shall go: in taking a solitary ramble, the question is what we shall meet with by the way. The mind then is "its own place"...' (p. 145). This allusion to Milton's Satan, however perfunctory it may be, nonetheless should make us pause for thought, since it does alert us to some possible ambivalences in the essay. If Satan's speech taking possession of hell stands, for Romantic readers of Milton, as a triumphant assertion of the self's victory over circumstance, Hazlitt's essay, vigorously upbeat though it may be, partakes of a siege mentality that never fully convinces one that solitude is as life-affirming as it is said to be. In the final part of his essay he enumerates the situations in which companionship is welcome on a walk: these include excursions with a definite end in view ('ruins, aqueducts, pictures'), and all foreign travel. Surprisingly, Hazlitt's silent luxuriating in the unchallenged sway of his own sensations and observations, on a solitary ramble, yields when abroad to an appetite for company and communication: 'In such situations, so opposite to all one's ordinary train of ideas, one seems a species by one's-self, a limb torn off from society, unless one can meet with instant fellowship and support' (p. 146). Having 'to unravel [the] mystery of our being at every turn' (p. 139), which Hazlitt disdains when walking in home surroundings, becomes contrarily urgent when such 'mystery' is linked to a myth of national identity. This over-anxious sense of linguistic and cultural belonging, or wholeness, is seen as the precipitate of a post-Revolutionary and post-Napoleonic insularity: Hazlitt compares the joyous internationalism of the early years of the Revolution, when he first set foot 'on the laughing shores of France' and was unconscious of any 'alien sound' because he 'breathed the air of general humanity' (p. 146), to the post-war rediscovery of the prison-house of nationality.

Yet, as Hazlitt concludes his essay, another dimension to his new-found uneasiness with unaccompanied foreign travel comes into view. It is not so much the lack of an interlocutor who shares a language and a frame of reference that is troubling, but the estrangement from one's *internal* interlocutor:

> Our romantic and itinerant character is not to be domesticated. Dr Johnson remarked how little foreign travel added to the facilities of conversation in those who had been abroad. In fact, the time we have spent there is both delightful and in one sense instructive; but it appears to be cut out of our substantial, downright existence, and never to join kindly on to it. We are not the same, but another, and perhaps more enviable individual, all the time we are out of our own country. We are lost to ourselves, as well as to our friends. (p. 147)

We see here, rather poignantly, the separating-out of a Romantic self, an 'ideal identity', from an ordinary, actual, 'domesticated' self. The former, self-expansively, takes pleasure in the strange, the exotic, the unassimilable, but is itself unassimilable to the homely contours of one's everyday, waking personality. There seems to be more at stake here than foreign travel, more of a general disabused reflection on the fate of the Romantic Ego. When Hazlitt says that 'we can be said only to fulfil our destiny in the place that gave us birth' (p. 147), the last phrase has the sense of Wordsworth's 'world which is the world / Of all of us', where 'We find our happiness, or not at all' (*Prelude*, X.725–7), and carries the same counter-idealising message as that passage does in context.

So Hazlitt's late-Romantic walker is one who walks to evacuate and free the self from its regular social co-ordinates, but not to stray so far, literally or metaphorically, that that society is not at hand to perform a necessary infilling if the psychological anchors wear loose.

In its self-consuming meditation on the relative claims of solitary and companionable walking, 'On Going a Journey' focuses a prominent concern in peripatetic writing of the later Romantic period. In much of the tour literature, and a lot of the other peripatetic verse and prose, that I have discussed so far, companions are shadowy, even rhetorical, presences, who play no part in the discursive economy of the text. (There are, of course, significant exceptions: Dorothy Wordsworth's travel journals present a record of the social interactions of an extended family in transit.) In a range of texts by 'second generation' Romantics that I want to consider now, the companion (not accidentally, the title of one of Leigh Hunt's periodicals) moves to centre-stage, either (as in Hazlitt's essay) as an object of study for peripatetic theory or as a dialogised other in narratives of travel. The issue of walking *versus* talking, and of the different activities of mind which these terms connote (self-communing, reverie, aesthetic play *versus* performative language and conversational give-and-take) thus acquires new prominence. Another issue, on the edge of Hazlitt's essay, which I want to explore in this chapter, is the relation of rural to urban walking. Hazlitt leaves his reader in no doubt that the undisturbed freedom of contemplation he seeks can only be enjoyed when he absents himself from the town, and implies that 'Our romantic and itinerant character' (p. 147) can be experienced only in an exurban setting; but the late Romantic period also sees an intensification of literary interest in city walking, in the peripatetic encounter with new urban realities, that was to reach its apogee in the works of the Victorians and Modernists.

However, I want next to look closely at one final pedestrian tour on a traditional picturesque itinerary: John Keats's tour of the Lake

District and the Scottish Highlands in June-August 1818, in the company of Charles Armitage Brown. This trip has a certain notoriety in Keats criticism, because it seems that the physical exertions and climatic rigours faced by the tourists had a determining impact on his health, leaving him vulnerable to the tuberculosis that would later kill him and his two brothers. Keats himself saw the trip as part of the preparation for his poetic career, describing it in a letter to Benjamin Haydon in April as 'a sort of Prologue to the Life I intend to pursue – that is to write, to study and to see all Europe at the lowest expence'. He adds that he will 'get such an accumulation of stupendous recollolections [sic] that as I walk through the suburbs of London I may not see them'.[2] This cheery prediction suggests a superiority of the imagination to a drab urban environment that Hazlitt declared himself unequal to achieving, but the idea of mental accumulation has similarities with Hazlitt's advocacy of the 'synthetical method' on a journey, whereby he aims 'to lay in a stock of ideas then, and...examine and anatomise them afterwards' (p. 138). Stuart Sperry has offered a more elaborate and subtler account of the role played by the walking tour in Keats's aesthetic education: focusing on the visits to Robert Burns's tomb and birthplace, he analyses Keats's attempts to reconcile his respect for literality and historical truth with 'the etherealizing power of the imagination', which not only tended to apply false standards to the landscape but also threatened to commit violence on Burns's memory. Expanding on the conventional view that derives the settings of *Hyperion* (written in the autumn of 1818) from the mountain scenery Keats experienced for the first time on his walk, Sperry concludes that the tour 'forced him to bridge the gap between the grander forms of landscape and an ideal of the heroic in human nature in a way necessary for an epic attempt beyond the merely perfunctory or banal'.[3]

However, I do not want to pursue the question of the overall biographical significance of the northern tour, or of its contribution to furthering Keats's enquiries into beauty, the imagination, and the 'burden of the Mystery'. Although an approach which subordinates the literary output of the tour to the larger map of his life and career may reflect a fair estimate of its objective worth, I want to follow the reverse strategy of assessing how Keats's background, circumstances and rich talents allow him to make a unique intervention in the (by now) heavily overdetermined field of Romantic travel writing. In so doing, I do not want to minimise the fact that this intervention largely took the form of letters, and that some of the qualities I comment on would never be found in conventional tour writing for reasons of genre propriety; on the contrary, I want constantly to keep in mind that Keats, unlike Brown, channelled all his energy into this kind of literary record, purposely eschewing the more formal written tour which he derided by analogy with the Laputan printing press.[4]

The issue of sociability is a useful place to start. Brown is no invisible companion on Keats's tour. He was eight years older than Keats, by far the senior traveller (having run an office in St Petersburg as part of an export business owned with his brother), and was responsible for instigating and planning the tour. When Keats was forced to abandon the walk at Inverness and return to London, Brown continued alone. He and Keats, both resident in Hampstead, had met a year earlier, and their friendship appears not only to have survived the intimacy of the tour, but to have deepened: Brown took Keats under his wing after the death of his brother, nursed him for part of his final illness, and helped to advance his posthumous reputation.[5] On the evidence of his letters, Keats was far from agreeing with Hazlitt that he could not see the 'wit' of walking and talking at the same time; on the contrary, he creates wit from the shared experience of walking and talking. He begins a letter to his brother Tom in mid-July with forty lines of a mock Galloway song which, he says, 'Brown wanted to impose... upon dilke' (with whom Brown shared a large two-family home; *LJK*, p. 125). His next letter avails itself of some bawdy wordplay on the King Arthur legend attributed to Brown:

> Here's Brown going on so that I cannot bring to Mind how the two last days have vanished – for example he says 'The Lady of the Lake went to Rock herself to sleep on Arthur's seat and the Lord of the Isles coming to Press a Piece and seeing her Assleap remembered their last meeting at Cony stone Water so touching her with one hand on the Vallis Lucis while [t]he other un-Derwent here Whitehaven, Ireby stifled here clack man on, that he might her Anglesea and give her a Buchanan and said.' (*LJK*, p. 129)

Anyone needing help with the sexual puns here can consult Robert Gittings's helpful glossary.[6] I quote the passage in order to make the point that the friendship between Keats and Brown – evident through disagreement on moral/social issues,[7] and exchange of opinion on literary interests, as well as through humorous banter and practical joking – is woven prominently into Keats's letters, and that the ease of communicating the relaxed sociability of this male relationship[8] in letter form seems important to Keats's construction of a foliar, noncohesive 'Book of Travels' (*LJK*, p. 125).

This element of dialogic interaction is compounded and diversified by the array of correspondents addressed by Keats's tour literature, and the variety of epistolary contexts and motives through which he processes his narrative. The surviving letters involve six different addressees, and Keats is hypersensitive to their individual needs:

> I wish I knew always the humour my friends would be in at opening a letter of mine, to suit it to them nearly as possible I could always find an egg shell for Melancholy – and as for Merriment a Witty humour will turn any thing to Account – my head is sometimes in such a whirl in considering the million likings and antipathies of our Moments – that I can get into no settled strain in my Letters (*LJK*, p. 121)

Unlike Coleridge on his tour of Wales in 1794, who economises by including much of the same reportage, and many of the same jokes, in a letter to Henry Martin as in his earlier letter to Southey, Keats's letters are bespoke productions. In terms of what Jerome McGann would call their 'bibliographical codes',[9] Carol Kyros Walker comments that the letters are 'works of graphic art' and 'distinct artefacts', ranging from a text 'in a relaxed hand ... with an open, spatially liberal expansion on the page', to sheets bearing 'tight grids' of writing in both horizontal and vertical directions.[10] In terms of his awareness of audience, he tells his brother Tom in late June to 'Let any of my friends see my letters', and says that he is 'Content that probably three or four pair of eyes whose owners I am rather partial to will run over these lines' (*LJK*, p. 103); but it is as though he anticipated that one day they would be collected and bolted together as a continuous travel text. But the high degree of individuation manifested by the letters is enhanced by Keats's continuing immersion in a variety of emotional scripts. When writing to George and Georgiana Keats (his brother and sister-in-law), from whom he had just parted at Liverpool as they prepared to emigrate to America, Keats interrupts his travelogue to reassure them as to their fitness for the journey and throws off an acrostic on Georgiana's name to keep their spirits up. When writing to the poet and critic John Hamilton Reynolds, he halts his animadversions on Burns's infidelities to assure his correspondent that his pessimism is not complete on this issue, and that 'one of the first pleasures I look to is your happy Marriage' (*LJK*, p. 123). When writing to Bailey, he takes time to make fulsome apology for some mysterious offence he has caused, points out that he is following Bailey's advice to read Dante ('the only Books I have with me are those three little Volumes' [*LJK*, p. 138]), and juxtaposes mention of the kindness of a young woman who 'walked a Mile in a missling rain' to put him on the right path, with gratitude for some kind words of Bailey's about his sister. However routine these transactions of family and friendship may be, it is still worth remarking the extent to which Keats makes his chief literary record of the tour that was to be the prologue to his life of writing – the *Prelude* to his *Recluse* – an intersubjective event.

A new perspective can be gained on this matter by linking it to Keats's association with the so-called 'Cockney School' of literature grouped around Leigh Hunt, which received its patronising soubriquet from a series of aggressive and anonymous reviews in *Blackwood's Magazine* in 1817–18.[11] The fourth review in the series, which took Keats's 1817 *Poems* and *Endymion* as its subject, in fact appeared whilst he was on the walking tour, but the earlier reviews had already identified him as one of the School and he knew that he was in the firing line.

Jeffrey N. Cox has reassessed the literary output and relations of the Cockney School and made a strong case for raising the critical profile of such coteries, representing as they do 'the living ground upon which individuals come to share in a more widely diffused ideology'.[12] As well as constituting a loose, shifting, yet quite large and very active convivial group, the Cockney School expressed its solidarity in coterie publications such as Hunt's *Foliage* (1818), with its many poems addressed to members of the circle. For Cox, Romantic criticism has tried too hard to detach Keats, Shelley and Byron from this cultural context to secure their autonomy as great poets; to different extents, but Keats especially, they have to be seen as part of a liberal 'intelligentsia' with the *Examiner* as their central platform, and with 'reform, anticlericalism, and joyful paganism' as their creed.[13] The coterie quality of the School's literary production is evident in the popularity of occasional verse, in the presence of much interactive or collaborative work, and in the fondness for circulating material in manuscript. Cox's most original contribution comes in relating the terms of the *Blackwood's* attacks to contemporary stereotypes of the Cockney character (wanton, effeminate) and current definitions of Cockney culture (a site of lively popular rituals, including the annual mock-subversive crowning of a Cockney King of London). What was seen by members of the School as a 'programme of cheerfulness and sociality' to set against the 'self-centered, "money-getting" spirit of their day', was likely to be misprised by conservative critics as 'affectation and conceit', 'swagger' and 'insolence'. The latter were right, Cox argues, in identifying a style that was indecorous according to the dominant aesthetic ideology of the day; it was a style which, with 'its diction shifts, its "new-fangled" feel, its odd juxtapositions...was an attempt to capture the pulse of modern city life'.[14] Further detail on the lineaments of this style can be obtained from William Keach's work on Keats's versification: taking his cue from *Blackwood's* derision of Keats's 'Cockney rhymes', Keach explores the analogy between the poet's 'loose liberal couplets' (freely enjambed) and the 'loose liberal politics he had imbibed from Hunt', pointing out that even Hunt accused Keats of introducing 'roughness and discord for their own sake, but not for that of variety and contrasted harmony'.[15]

If this revisionist view of the importance of Keats's membership of the Cockney School is to be trusted, it gives new significance to the intersubjective cast of his travel writing of 1818, since this can be seen as the continuation in another genre of that 'programme of cheerfulness and sociality' which had so far governed his literary life under the wing of Hunt. Keats's aspersions on the Scottish Presbyterian church, pronounced from the vantage-point of his brief stay in Ireland, are also relevant in this connection. Although his horror at the rural living conditions he sees in the country around Belfast leads him to acknowledge the benefits of the Kirk's turning the Scots into 'regular Phalanges of savers and gainers', his temperamental and ideological aversion to its dour morality cannot be repressed:

> These kirkmen have done Scotland harm – they have banished puns and laughing and kissing (except in cases where the very danger and crime must make it very fine and gustful. I shall

make a full stop at kissing for after that there should be a better parent-thesis.... (*LJK*, p. 118)

It is characteristic of Keats's writing that his word-play on 'stop' and 'parenthesis' here performatively distances the author from the proscriptions he denounces. When he goes on to say that he would 'sooner be a wild deer than a Girl under the dominion of the kirk' (*LJK*, p. 118), one recognises a small outcrop of that 'pagan' instinctuality which Jeffrey Cox and Marilyn Butler have pointed to as part of the Cockney credo.[16]

Moving beyond this general equation of *Examiner* morality and the ethos of Keats's pedestrian tour, it is possible to discern other elements of Cockney 'swagger' and good humour in his letters on the journey. For a start, there is his inability to maintain the 'correct' aesthetic stance. Despite his evident delight at the Ambleside waterfall, which astonishes him with its 'tone' and colouring, and via which he furthers his speculations on the relative values of perception and imagination, he cannot resist noting its realistically small dimensions in the diminishing metaphor of a 'teapot spout' (*LJK*, p. 104). At Lodore Falls he admires the picturesque qualities of the scene, but punctures the effect by recording how he got 'damped by slipping one leg into a squashy hole' (*LJK*, p. 107). The description of Fingal's Cave is conventionally elevated in tone (however fresh some of the images), but his account of the ascent of Ben Nevis is more characteristically particoloured. He amply acknowledges the physical demands of the climb, and his wonder at contemplating the deep rifts in the side of the mountain, and the longitudinal perspectives from the summit, is intense and genuine; but there is an undertow of (self-)ridicule and comic embellishment throughout which unsettles, and finally breaches, literary decorum. He is 'a fly crawling up a wainscoat' (*LJK*, p. 145); the mountaintop changes its weather more often than a lady her head-dress; and there is a pantomime *reductio ad absurdum* of their negotiation of the loose stones near the summit:

> sometimes on two sometimes on three, sometimes four legs – sometimes two and stick, sometimes three and stick, then four again, then two[,] then a jump, so that we kept on ringing changes on foot, hand, Stick, jump boggl[e,] s[t]umble, foot, hand, foot, (very gingerly) stick again, and then again a game at all fours. (*LJK*, p. 147)

Finally, Keats transforms the aesthetic code of his narrative by inserting a bawdy piece of verse drama about a querulous Mrs Cameron who nearly falls prey to the sexual appetite of the mountain: 'I shall kiss and snub / And press my dainty morsel to my breast'.[17] It seems questionable whether the experience of climbing Ben Nevis will fulfil Keats's intention (as a rationale of the tour) to 'load me with grander Mountains, and strengthen more my reach in Poetry' (*LJK*, p. 137); indeed, it is the satirical overtones of 'load me...with mountains' that are amplified and overrule the normal discipline of travel writing. In his final letter of the tour, to the mother of his sister-in-law, the remark that he has been '*werry* romantic indeed, among these Mountains & Lakes' (*LJK*, p. 150) sums up his semi-detached relation to that literary tradition.

This kind of double-voiced discourse, alternately straight and ironic, possessed of a 'Cockney' wit and impertinence, can also be considered in the light of Mikhail Bakhtin's well-known work on the semiotics of parody. For Bakhtin, the world of dialogised heteroglossia he sees as the greatest achievement of the modern novel has its roots in the multifarious parodic-travestying genres of classical and medieval literature. As regards classical culture, he speculates that there was never a straightforward genre that did not have its answering parodic double; as regards the Middle Ages, he stresses the strong links of parody with popular ritual, with the licensed freedoms of holidays and festivals – it is this aspect of his work that has led to the popularisation of 'carnivalesque' as a term signalling the inversion of hierarchies. In whatever context, parody performs the same function at an individual and social level:

> It is as if such mimicry rips the word away from its object, disunifies the two, shows that a given straightforward generic word – epic or tragic – is one-sided, bounded, incapable of exhausting the object.... Parodic-travestying literature introduces the permanent corrective of laughter, of a critique on the one-sided seriousness of the lofty direct word, the corrective of reality that is always richer, more fundamental and most importantly *too contradictory and heteroglot* to be fit into a high and straightforward genre.[18]

Furthermore, corrective laughter, whilst its permissive status may have denied it any real subversive potential, had the effect of liberating consciousness from the prison-house of a particular, hegemonic discourse, and therefore prepared the ground for the kind of relativised understanding of reality in which change might be possible.

If I am right in detecting the deliberate intrusion of a vein of 'Cockney carnivalesque' into Keats's pedestrian travel writing,[19] then *Blackwood's* was entirely mistaken in mocking him for not having 'learning enough to distinguish between the written language of Englishmen and the spoken jargon of Cockneys'.[20] On the contrary, Keats was not only well aware of the differences between the polite language in which most published tours were written and the 'spoken jargon of Cockneys', but took pleasure in dialogising the contrasting idioms and styles. Whereas even the best modern criticism can construe this as a weakness – Stuart Sperry talks of the intrusion of 'inappropriate details' and the 'tenuousness and instability of the marvelous' in one of the travel letters as 'old, recurring problems'[21] – Keats arguably took this course out of respect for the contradictory social reality in which he was placed.

One can go further in analysing the parodic-travestying aspects of Keats's travel literature. The jokes, word-play and vulgarity that permeate the letter-sequence, and which are an important part of the 'corrective of laughter', have already been illustrated: the acrostic on Georgiana Keats, the pun on 'parenthesis', the cod Galloway ballad, the masculinist humour of the Arthurian burlesque, and the poem on Mrs Cameron's ordeal on Ben Nevis, all exemplify this vigorous strain in Keats's writing. Another important feature of Bakhtin's theory of the carnivalesque is the emphasis on the grotesqueries of the human body and human behaviour – gluttony and drunkenness are sample motifs which show the link with holidays and festivals. This aspect of literary representation can also be illustrated from Keats's letters – where, as Carol Kyros Walker observes, the portraiture 'is virtually Chaucerian'.[22] One example would be the landlord at Burns's cottage (converted into a whiskyshop by the time of Keats's visit):

> The Man at the Cottage was a great Bore with his Anecdotes – I hate the rascal – his Life consists in fuz, fuzzy, fuzziest – He drinks glasses five for the Quarter and twelve for the hour, – he is a mahogany faced old Jackass who knew Burns – He ought to be kicked for having spoken to him. He calls himself 'a curious old Bitch' – but he is a flat old Dog – I should like to

employ Caliph Vatheck to kick him – O the flummery of a birth place! (*LJK*, p. 122)

The trip to Burns's cottage, given Keats's strong identification with Burns both socially and aesthetically, was to have been one of the highpoints of the tour; but the visit was a disillusioning one. In some interesting heptameter couplets despatched in a slightly later letter, this disillusionment is expressed through an extended metaphor of pilgrimage: the physical and mental commitment intensifies the pleasure of arrival, but is also isolating and delusion-building, and 'a longer stay / Would bar return and make a Man forget his mortal way' (*LJK*, p. 139). In the passage quoted, disillusionment is expressed as a vulgar dissonance, in the heteroglossic challenge of the drunken landlord's 'gab' which 'hindered my sublimity' (*LJK*, p. 122). The irony is that Keats himself abandons the high linguistic ground to give vent to his disgust and frustration.

A second example of the violation of decorum by the representation of degraded human bodies would be the following portrait:

> On our return from Belfast we met a Sadan – the Duchess of Dunghill – It is no laughing matter tho – Imagine the worst dog kennel you ever saw placed upon two poles from a mouldy fencing – In such a wretched thing sat a squalid old Woman squat like an ape half starved from a scarcity of Buiscuit in its passage from Madagascar to the cape, – with a pipe in her mouth and looking out with a round-eyed skinny lidded, inanity – with a sort of horizontal idiotic movement of her head.... (*LJK*, p. 120)

Whereas drunkenness and logorrhea were the forms of excess on display at Burns's cottage, here the exaggeration is entirely in the service of physical repulsion, the spectacle a human bestiary stripped of any moralising function: Keats marvels at a possible 'history of her Life and sensations', but makes no imaginative effort to research it.

The reader approaching Keats's travel literature for the first time may well be disappointed on discovering the nature of his poetic output on the tour. Certainly, he produced some notable sonnets: 'On Visiting the Tomb of Burns', 'To Ailsa Rock', and – undoubtedly the finest – 'Read Me a Lesson, Muse'. The remainder of his output, however, consists chiefly of ballads and satiric-comic verse. It perhaps becomes less obligatory to make excuses for this performance when one positions the poetry in the aesthetic context of the Cockney School and the culture wars of post-Napoleonic Britain, for these poems, however lightweight by standard Canonical measures, can be seen to play their part in directing a critique of laughter at the literary pretensions and hypocritical morality of the dominant discourse. I shall conclude my discussion of Keats by quoting from a song which Keats thought innocent enough to enclose in a letter to his sister, and which he said was about himself. I need to quote the whole of the final 'stanza':

> There was a naughty Boy
>   And a naughty Boy was he
> He ran away to Scotland
>   The people for to see –
>   There he found
>   That the ground
>   Was as hard
>   That a yard
>   Was as long,
>   That a song
>   Was as merry,
>   That a cherry
>   Was as red –
>   That lead
>   Was as weighty
>   That fourscore
>   Was as eighty
>   That a door
>   Was as wooden
>   As in england –
>   So he stood in
>   His shoes
>   And he wonderd
>   He wonderd
>   He stood in his
>   Shoes and he wonder'd –   (*LJK*, p. 115)

This poem might well be seen as the comic-decadent *reductio ad absurdum* of Keats's 'loose, nerveless versification, and Cockney rhymes'.²³ Here the prosodic quirk concerns not just the 'liberal' use of enjambement, but also the syncopated rhythm which denies closure at the end of each dimeter couplet, since the metrical and semantic units are consistently out of step with each other. William Keach quotes an interesting passage from Francis Jeffrey's review of *Endymion* in *The Edinburgh Review*, which accuses Keats of having 'taken the first word that presented itself to make up a rhyme, and then made that word the germ of a new cluster of images – a hint for a new excursion of the fancy – and so wandered on, equally forgetful whence he came, and heedless whither he was going...'.²⁴ This comment, with its uncannily apt pedestrian metaphor, seems uniquely fitted to Keats's 'naughty Boy' song, whose elongated stanzas make a cheerily opportunistic progress down the page. As for its theme, the notion of the boy running away to Scotland and finding everything much the same as in England is a further instance of Keats's bantering irreverence towards the institution of 'scenery and visitings' and the kinds of 'official' literature it sponsored.

By way of summary, there is an episode in Charles Brown's account of the tour, the serialised 'Walks in the North', describing an encounter in Ambleside with an Oxford graduate excessively conscious of the image he is presenting, and making constant reference to fashionable life in London. Brown comments that Ambleside is 'an unfit spot... for the speaking of a particular town-bred impertinence'.²⁵ My reading of Keats's letters has suggested that he himself filters his travel experience through *another* kind of 'town-bred impertinence'. If the Cockney style was an attempt to capture the pulse of city life, as Jeffrey Cox argues, then Keats's travel writing might be taken to depict an urbanising excursion into the wilder parts of the country, an ironic counterpart to the 'ruralizing imagination' at work in much literature of the city.²⁶ It is urban peripatetic that will concern me for the remainder of this chapter.

Raymond Williams observes in passing, in *The Country and the City*, that 'perception of the new qualities of the modern city had been associated, from the beginning, with a man walking, as if alone, in its streets'.²⁷ He traces this beginning to Blake's 'London', but recognition of the contradictory realities of the rapidly-expanding capital city went back much earlier, as Williams is well aware. He himself points to the contrast between the urbanity of Pope, Johnson and Swift, and the blighted vision of Hogarth, Fielding, Gay and Defoe. Equally, the figure of the walker, as peripatetic observer or amateur anthropologist, is on hand well before the Romantic period to relay these contradictions. Swift's 'Description of a City Shower' (1710) offers its view of a London crowd, at once thrown into confusion and brought to one level, from the standpoint of a pedestrian poet bespattered along with everyone else. John Gay's *Trivia; or, the Art of Walking the Streets of London* (1716) gives, through the filters of high burlesque, a fuller impression of ways of seeing the city at the beginning of the eighteenth century. For example, although the fashionable West End was still hemmed in by open countryside at this time, it is possible for Gay, perambulating the streets at Seven Dials, to see London as a perilous labyrinth:

> Here oft the peasant, with enquiring face,
> Bewilder'd, trudges on from place to place;
> He dwells on ev'ry sign with stupid gaze,
> Enters the narrow alley's doubtful maze,
> Tries ev'ry winding court and street in vain,
> And doubles o'er his weary steps again.   (II. 77–82)[28]

In fact, the entire structure of *Trivia* is one of variations on the theme of the city as assault course, confronting the walker with an array of physical and moral threats: by day, one has to protect one's clothing from soiling by miscellaneous tradesmen, one's money from pickpockets posing as guides, and one's head from falling tiles and masonry; by night, beggars turn into thieves, carriage traffic is doubly dangerous, and prostitutes lie in ambush for 'strangers unsuspecting hearts' (III. 264). Although Gay puts forward positive arguments for the 'happiness' of walkers on health and safety grounds, the poem's keynote is the unceasing vigilance required of the traveller. We are some way here from the 'experience of shock' which Walter Benjamin, in his reading of Baudelaire, saw as the prototypical response to the modern city, yet the anxiety with which, beneath the mock-heroic brio, Gay's walker seeks to steer a course through the crowd, maintaining a protective distance from the denizens of the nether world, anticipates in certain respects the psycho-social conditions of urban existence that Benjamin describes.[29]

The more positive reading of the city in *Trivia* is of an agreeable flow of superficial appearances – especially agreeable if they can be enjoyed in congenial company – , and of a bountiful centre of trade and urbane shoppers' paradise. In Wordsworth's well-known account of his residence in London in Book VII of *The Prelude*, these same features, of a city now considerably expanded, recur initially as the source of pleasurable distraction:

> The broad high-way appearance, as it strikes
> On Strangers of all ages, the quick dance
> Of colours, lights and forms, the Babel din
> The endless stream of men, and moving things,
> From hour to hour the illimitable walk
> Still among streets with clouds and sky above,
> The wealth, the bustle and the eagerness,
> The glittering Chariots with their pamper'd Steeds,
> Stalls, Barrows, Porters.... (VII. 155–63)

Eventually, Wordsworth is forced to denounce this same bewitching phantasmagoria, which is placed in a savagely distorting mirror as the 'Parliament of Monsters' that is Bartholomew Fair (VII. 644–94). The catalyst for this countervaluation is a growing sense of incompatibility between 'the press / Of self-destroying, transitory things' (VII. 738–9) and the idealising imagination, and a revulsion from the impersonality and anonymity of the city. Raymond Williams identifies Wordsworth's account of this traumatisation, which culminates in the famous image of the Blind Beggar, identity-tagged, 'a type...of the utmost that we know' (VII. 617–8), as the first literary expression of urban alienation.[30] This trauma, it is worth noting, happens to one who has 'gone forward *with* the Crowd' (VII. 596; my emphasis), who has surrendered his mental privacy and independent will to the life of the street.

The street as a place of alienated and alienating encounters is therefore central to the experience of the Romantic city-walker. One image that sticks in the mind is that of William Hazlitt's devastating chance encounter, in the 'desert streets' of London, with the double-dealing object of his sexual obsession, Sarah Walker (his landlady's daughter). At a loss to understand Sarah's enigmatic behaviour, Hazlitt is finally relieved of all doubt when he sees her in the company of another man:

> We passed at the crossing of the street without speaking. Will you believe it, after all that had passed between us for two years, after what had passed in the last half-year, after what had passed that very morning, she went by me without even changing countenance, without expressing the slightest emotion, without betraying either shame or pity or remorse....[31]

Despite the special circumstances, the slightly surreal quality of this meeting seems paradigmatic of the estranging effects attributed by many Romantic and post-Romantic writers to the exterior life of the city. However, just as with Gay a hundred years earlier, alternative responses were available, as Williams proves in underlining the ambivalence of Wordsworth's concluding judgement, with its 'historically liberating insight' of 'new kinds of possible order, new kinds of human unity'.[32]

One interesting text embodying these complex perceptions of the city is De Quincey's *Confessions of an English Opium-Eater*. The first half of this autobiography gives a gripping account of the author's teenage life as a self-made vagrant, from his nocturnal flight from Manchester Grammar School, to his itinerant existence in North Wales, to his arrival in London, where he suffers great hunger and for two months 'very seldom slept under a roof',[33] unwilling to contact family friends for fear of being returned to school by his guardians. During this period he becomes a compulsive 'peripatetic', mingling with the 'poor houseless wanderers' of the capital and befriending a prostitute called Ann, with whom he walks at night and sleeps in doorways (p. 50). However, the very term 'peripatetic' signals the in-betweenness of De Quincey's position, enduring malnutrition and dangerous ill-health yet able at any time to rescue his respectability; and, indeed, he makes a virtue out of necessity by stressing the value of his intimacy with the London underclass to the disinterested vision of the philosopher, who should stand 'in an equal relation to high and low – to educated and uneducated, to the guilty and the innocent' (p. 50). De Quincey is neither the man of leisure, the *flâneur*, nor Poe's 'Man of the Crowd', who refuses to be alone and craves the inarticulate solidarity of pedestrian traffic. He is intellectually divorced from the houseless poor, yet he finds the street a site of possible connections. Ann is the major exception to his 'philosophical' detachment, leading him, when he inadvertently loses contact with her, to frenzied searches through the 'mighty labyrinths' of

London (p. 64); but he also describes opiated Saturday-night wanderings in the poorer quarters of the capital, observing yet also conversing with people, sympathising with their pleasures and disappointments. When he gets lost in these perambulations, the opium has the dual effect of intensifying the sense of labyrinthine mystery, and of transforming the 'perplexities of my steps' (p. 81) into moral and intellectual confusion – a sublimation of physical plight into mental disturbance which has lasting power over him.

Benjamin writes that Baudelaire remained conscious of the social realities of Paris only in the way that intoxicated people are still aware of reality, a condition prefigured in De Quincey's account of his opiated perceptions of the 'sphinx's riddles' of London streets. Benjamin's highly speculative remarks on the analogy – grounded in the idea of intoxication – between people and commodities also throws light on a significant (unwitting?) play on words in the *Confessions*, where in the same sentence he refers to himself as a 'walker of the streets' and to prostitutes like Ann as 'street-walkers' (p. 50). For Benjamin, the stroller shares the nature of the commodity, surrounded by a surge of 'customers' whom he imaginatively enters only if they are worth inspecting:

> If the soul of the commodity which Marx occasionally mentions in jest existed, it would be the most empathetic ever encountered in the realm of souls, for it would have to see in everyone the buyer in whose hand and house it wants to nestle. Empathy is the nature of the intoxication to which the *flâneur* abandons himself in the crowd.[34]

The prostitute has explicitly commodified the human body, but the *flâneur*, through the peddling of his imaginative sympathies, is complicit in what Benjamin terms 'the prostitution of the commodity-soul'. In this light, De Quincey's peripateticism is not as free of 'impurity' as he would like to think: he is as much a 'street-walker' as Ann.

Perhaps a more representative, and certainly less tormented, Romantic *flâneur* is Leigh Hunt, and by way of circling back to the cultural values of the Cockney School I shall finish with some observations on the contribution made to the literature of walking by its leading light. If, as Benjamin says, the *flâneur* stands midway between a world of complete leisure and the turmoil of the city,[35] then Hunt is that individual: he can identify completely neither with the rural or suburban calm he sentimentalises, nor with the populous commercial centre where he made his living. The keynote of a number of Hunt's essays is a good-humoured, sociable pedestrianism, expatiating on objects of interest of any kind. This could be seen, perhaps, as an attenuation of the Romantic ideology of freedom-through-walking, and its replacement with a lighter, recreational-touristic creed. Whereas Hazlitt, when going a journey, puts the town both out of sight and out of mind, the urban/rural divide is smoothed over in Hunt's work: 'A Walk from Dulwich to Brockham' (1828) is a typical piece of littoral walking, valuing 'remoteness' only in terms of 'how far you feel yourself from your commonplaces'.[36] In the heterogeneity of his concerns, which range here from conventional tributes to pastoral beauty to remarks on church architecture or the desirability of diffusing knowledge among the poor, Hunt never seems anything but a townsman abroad. 'Of the Sight of Shops' (1828) weighs up the relative merits of rural and city walking, and suggests that a peopled environment is sometimes preferable: 'If you have been solitary... for a long time, it is pleasant to get among your fellow-creatures again, even to be jostled or elbowed.'[37] In later essays he confines himself to perambulations in London: his 'Ramble in Mary-le-Bone', published in *The Weekly True Sun* in 1833, is an urbane piece of cultural topography with a slightly hardened edge, his commentary on old buildings and their eminent inhabitants maintaining a mocking, sceptical tone towards royalty and aristocracy. In the essays collected in *A Saunter in the West End* (1861), his rambling is entirely dephysicalised: although he addresses his readers 'as if they were lovers of local associations walking along the pavement at their leisure', the impression is that, rather in the manner of a royal visit, the streets have been cleared to enable Hunt to take his (mental) walk in peace.[38]

Like Gay, and like Wordsworth on his first exposure to London, it is the mutable exterior form of the city which captivates Hunt as a pedestrian. Showery weather can accentuate the delights of 'a quickly-dried pavement and a set of brilliant shops'; 'a run of agreeable faces' can lend a street a finer aspect than the country; and the shops which appeal most to the easily-distractable imagination are those which suggest ideas that are 'obvious and on the surface'.[39] The material displeasures of the city (to which Gay was so attentive) can merely enhance appreciation of its peculiar comforts: the picture he constructs in 'Bad Weather', of a 'narrow, foggy, and noisy' street, in which 'the pedestrians kick, as they go, those detestable flakes of united snow and mud', is compensated by the sensuous enjoyments of a pampered domesticity.[40] Alternatively, the city-walker caught in a mini-dust storm caused by a carriage or a flock of sheep in a narrow street, in 'A Dusty Day', rises above his predicament by intellectualising his antagonist – whether by making epigrams on the experience or by meditating on particles of dust as the stuff of which the whole cosmos is made. Such reflections are 'the weapons with which [the pedestrian] triumphs over the most hostile of his clouds, whether material or metaphorical'.[41]

This last example might suggest that Hunt is as interested in arming the city-walker's consciousness against the experience of shock as in sensitising it to an emporium of fresh sensations. These tensions are certainly present. In 'A Nearer View of Some of the Shops', Hunt penetrates the interiors debarred to the companion essay, and takes particular pleasure in a sculptor's gallery, recalling how 'the world used to seem shut out from us the moment the street door was closed, and we began stepping down those long-carpeted aisles of pictures, with statues in the angles where they turned': the pulse of urban life is slowed in the presence of fine art.[42] 'Walks Home by Night', which describes a typical journey from the West End to his home in the leafy suburb of Highgate, is a more complex instance. What appeals most to Hunt about a nighttime walk is the abolition of all that characterises the city, the tranquillising of nervous humanity: 'Inanimate objects are no calmer than passions and cares now seem to be, all laid asleep. The human being is motionless as the house or the tree....'.[43] The response is comparable to Wordsworth's, in the best-known of all ruralised images of the city, the sonnet 'Composed Upon Westminster Bridge'. As Hunt proceeds, however, he begins to think into the situation of an apothecary, called out of bed to minister to some 'hurt mind'; but then he refuses to get 'too much into the interior of the houses', and, as if in parodic illustration of this retreat, escapes from a worrisome dog. Jeffrey Robinson's comment on this development is apt: 'The city, it seems, encourages the imagination to penetrate into it; it encourages the arousal and exercise of a sympathetic, social consciousness. Hunt...draws back from the city's encouragement.'[44] This would be unfair as an overall judgement on Hunt's work, but it is an accurate summary of this essay. The second half consists of recollections of watchmen he has encountered on his walks home. One such individual, a 'bale of a man' sliding extravagantly on the ice, had 'slipped from out of his box and his commonplaces at one rush of a merry thought, and seemed to say, "Everything's in imagination; – here goes the whole weight of my office"'. It seems a typically playful metaphor for Hunt's own withdrawal into his private self. As his walk ends, and he deposits himself in his 'nest', he seems only to have substituted one form of imitation 'nature' for another.

The complementary Huntian paradox of an urbanising incursion into the country is well illustrated by a late poem, which, because it marries many characteristics of peripatetic and the Cockney style, I shall use to bring down the curtain on the age of pedestrianism. 'A Rustic Walk and Dinner' was first published in *The Monthly Magazine* in 1842. It is written in 'intentionally unelevated' blank verse – 'literally', Hunt's preface says, *'sermo pedestris*, – poetry on foot';[45] as such, it continues the line of verse peripatetic represented by Cowper, Coleridge, Wordsworth and others. It begins thus:

> How fine to walk to dinner, not too far,
> Through a green country, on a summer's day,
> The dinner at an inn, the time our own,
> The roads not dusty, yet the fields not wet,
> The grass *lie-down-upónable*. (ll. 1–5)

The outrageous compound adjective shifts what is already colloquial ease into the 'bravado style' associated with the Cockney School.[46] Hunt continues by celebrating the pleasures of travelling on horseback, by coach, and by boat, but concludes that 'walking's freest' because 'you command / Time, place, caprice; may go on, or return, / Lie down, expatiate, wander' (ll. 33–7). Enlisting the reader as his ideal companion, he narrates one such walk to a country inn, passing through the suburbs on the roads,

> And then by field-paths, and more flowery ditch
> White-starred, and red, and azure, – and through all
> Those heaps of buttercups, that smear the land
> With splendour, nearly extinguishing the daisies, –
> And hill, and dale, and stile on which we sat
> Cooling our brows under the airy trees,
> And heard the brook low down, and found that hunk

Of bread so exquisite, to the very crumbs
That shared a pocket-corner with its halfpence. –
(O Shelley! 'twas a bond 'twixt thee and me,
That power to eat the sweet crust out of doors!
You laughed with loving eyes, wrinkled with mirth,
And cried, high breathing, 'What! can you do *that*?
I thought that no one dared a thing so strange
And primitive, but myself.' – And so we loved
Ever the more, and found our love increase
Most by such simple abidings with boy-wisdom.)   (ll. 56–72)

The subject-matter here is obstinately mundane, the syntax loose and irresolute, in sympathy with the notion of 'poetry on foot'; the long parenthesis, which beguilingly recalls the intimacy with a former associate member of the Cockney School, speaks more generally to the School's values of cheerful informality and democratic 'primitivism'.

Having already absorbed elements of conversational exchange, the poem now assumes explicitly dialogic form, as Hunt presents himself and the reader seated bird-like in a tree, engaging in unsolemn, irreverent philosophical discussion on the necessary imperfections of existence, and reconciling Christianity and paganism in the formula, 'God is "good", the gods but good divided' (l. 170). As they continue their walk, their attention is divided between the actual materials of perception (dry ditch, nettles, cuckoo, thrush) and the fancied shapes of myth, literature and legend (notably Robin Hood). Eventually, they look down from a hill onto the village where their inn is located, indulging in sentimental pastoralism, then ironising it forthwith: farms represent a homely golden age, 'did but the inmates know it' (l. 314). In the same way, in the following topical passage, it is Hunt's verbal cleverness that 'cuts through' the green pastures, as well as the railway:

Far be their rail-roads from this quiet spot,
Cutting its heart through; – far that *anti-farness*,
Trampling all peaceful places into forced
And iron neighbourhood; making all towns
O'ertake all country with their shoes of swiftness,
That stamp their tyrannous tracks in steel for ever,
Killing the green, the loneliness, the poetry.   (ll. 321–7)

A final tramp across fields individualised by reference to different meadow-flowers and hedgerows brings them to the inn, where they *'flop'* (l. 374) into their seats. There follows an extensive mock-heroic description of their well-earned meal ('Now sticketh fork in flesh, and the chops vanish' [l. 411]), which nevertheless establishes the superiority of their 'ill-bred smotherments of munch' (l. 417) to a formal dinner in polite company.

The poem ends in a benign, well-fed torpor of love and tolerance – the sort of easy-going goodwill that Hunt has communicated throughout. To the end, the poem exudes 'Cockney' wit, charm and swagger, a kind of adult delinquency in language and behaviour. *Contra* Hazlitt, it celebrates the social pleasures of walking and talking at the same time. It is completely likeable. It is also, it has to be said, somewhat inconsequential. But the context has altered: Cockney morality could not possess the same oppositional quality in the 1840s as it did in the post-war era; equally, the once radically levelling culture of pedestrianism had become assimilated to the mainstream. Only now, as we near the end of the twentieth century, have conditions been re-established in which walking has become, for some, a political act.

Notes

NOTES TO CHAPTER 1

1. Joseph Gilbert and Thomas Howell, *A Correct and Minute Journal of the Time Occupied in Every Mile By Mr. John Stokes, of Bristol, During his Walk of Fifty Miles per Day for Twenty successive Days, making One Thousand Miles, At Saltford, in the County of Somerset, from the 20th of November to the 9th of December 1815* (Bristol: C. Frost et al., 1815) pp. 12, 7, 10.
2. Walter Thom, *Pedestrianism: or, an Account of the Performances of Celebrated Pedestrians during the Last and Present Century* (London, 1813) p. 31. Among the earliest notable pedestrians Thom cites are 'Child of Wandsworth', who walked 44 miles in 7 hours 57 minutes on Wimbledon Common in 1762, and John Hague who did 100 miles in 23 hours 15 minutes in the same year.
3. Thom, *Pedestrianism*, p. 205.
4. Wilson was prevented from completing his walk of 1000 miles in 20 days, for a subscription of 100 guineas, by the local magistrates, who issued an injunction against him for breach of the peace and violating public morals. His own *Sketch of the Life of George Wilson, the Blackheath Pedestrian* (London: Printed for the Author, 1815), which contains an account of his chequered working life in shoemaking, pawnbroking and secondhand trading, a plangent defence of his sometimes violent relationship with his adulterous wife, and a sustained ironic denunciation of the Blackheath magistrates, defies paraphrase.
5. See the anonymous *Short Sketch of the Life of Mr. Foster Powell, The Great Pedestrian* (London: R.H. Westley, 1793). One of Powell's major performances was walking 100 miles on the Bath road in $23\frac{1}{4}$ hours in 1786, which 'caused much noise in *London*, and made his fame spread abroad as a Pedestrian' (p. 7).
6. Thom, *Pedestrianism*, pp. 41, iii–iv.
7. Morris Marples, *Shank's Pony: A Study of Walking* (London: Dent, 1959) pp. 31, 78.
8. Marples, *Shank's Pony*, pp. 31–3.
9. Anne D. Wallace, *Walking, Literature, and English Culture: The Origins and Uses of Peripatetic in the Nineteenth Century* (Oxford: Clarendon Press, 1993) pp. 8, 9.
10. Wallace, *Walking*, pp. 63, 54, 53.
11. William Coxe, *Travels in Switzerland and in the Country of the Grisons, in a Series of Letters to William Melmoth, Esq.*, 2 vols. (Paris: James Decker, 1802) vol. I, p. 1.

12. Coxe, *Travels*, vol. I, pp. 31–2.
13. William Lisle Bowles, 'Sonnet 2', in *Sonnets, Written Chiefly on Picturesque Subjects, During a Tour*, 2nd ed. (Bath: R. Cruttwell, 1789) p. 11. For information on Bowles's tour see Garland Greever (ed.), *A Wiltshire Parson and his Friends: The Correspondence of William Lisle Bowles* (London: Constable & Co., 1926).
14. Frida Knight, *University Rebel: The Life of William Frend (1757–1841)* (London: Victor Gollancz, 1971). Knight's biography contains an interesting account of Frend's tour.
15. Joshua Wilkinson, *The Wanderer; or Anecdotes and Incidents, The Results and Occurrences of a Ramble on Foot, Through France, Germany and Italy, in 1791 and 1793*, 2 vols (London: L.J. Higham, 1798) vol. I, p. 32.
16. John Stewart, *Opus Maximum: Or, the Great Essay to Reduce the Moral World from Contingency to System* (London: J. Ginger, 1803) p. xviii.
17. Michael Kelly, *Reminiscences of Michael Kelly*, 2 vols. (London: Henry Colburn, 1826) vol. I, pp. 247–8.
18. Anon., *The Life and Adventures of the Celebrated Walking Stewart: Including his Travels in the East Indies, Turkey, Germany, and America. By a Relative* (London: E. Wheatley, 1822) pp. 12–13.
19. Thomas De Quincey, 'Walking Stewart', *The Works of Thomas De Quincey*, 4th ed., 16 vols (Edinburgh: A. & C. Black, 1862–78) vol. VII, p. 1.
20. Letter to Dorothy Wordsworth, 6–12 September 1790, *Letters of William and Dorothy Wordsworth: The Early Years, 1787–1805*, ed. E. de Selincourt, 2nd ed., revised Chester L. Shaver (Oxford: Clarendon Press, 1967) p. 37. Cited henceforth as *EY*.
21. Adam Walker, *Remarks Made in a Tour from London to the Lakes of Westmoreland and Cumberland in the Summer of 1791* (London: G. Nicol & C. Dilly, 1792), Advertisement; Joseph Budworth, *A Fortnight's Ramble to the Lakes in Westmoreland, Lancashire, and Cumberland. By a Rambler* (London: Hookham & Carpenter, 1792) pp. xiii–xiv; Henry Skrine, *Three Successive Tours in the North of England, and Great Part of Scotland* (London: P. Elmsly, 1795).
22. J. Hucks, *A Pedestrian Tour Through North Wales, in a Series of Letters* (London, 1795; rpt. Cardiff: University of Wales Press, 1979) pp. 2, 5.
23. *Collected Letters of Samuel Taylor Coleridge*, ed. E.L. Griggs, 6 vols (Oxford: Clarendon Press, 1956–71) vol. I, p. 51. Henceforth cited as *CLSTC*.
24. Hucks, *Pedestrian Tour*, pp. 29, 5.
25. Anon., *The Observant Pedestrian; Or, Traits of the Heart: In a Solitary Tour from Caernarvon to London* (London; William Lane, 1795) pp. 52–7.
26. A.A., 'A Pedestrian Tour in North Wales', *Monthly Magazine*, I (1796) 15–18, 104–7, 191–4. Dyer's poem commemorating the latter tour was published in the *Monthly Magazine*, V (1798) 121–3.
27. A.B.L. Mandet de Penhouet, *Letters Describing a Tour through Part of South Wales. By a Pedestrian Traveller* (London, 1797) pp. i, 1.
28. Rev. Richard Warner, *A Walk through Wales, in August 1797* (Bath: R. Cruttwell, 1798), Advertisement.
29. Rev. W. Bingley, *A Tour Round North Wales, Performed during the Summer of 1798*, 2 vols. (London: E. Williams, 1800) vol. I, pp. 216–49, 375–82.

30. Unsigned review, *Monthly Magazine*, V (1798) 492.
31. Rev. R.H. Newell, *Letters on the Scenery of Wales; including a Series of Subjects for the Pencil, with their Stations Determined on a General Principle: and Instructions to Pedestrian Tourists* (London: Baldwin, Cradock & Joy, 1821) p. 2.
32. Newell's viewpoint receives support in Rosamond Bayne-Powell's study of travel in this period: 'It was supposed that no man of substance would ever walk, except with a gun over his shoulder, and that everyone who tramped the roads was either a footpad or a pauper. The roads were generally in such a bad state that walking could not have been pleasant and there was always the danger of attacks from footpads. It was not till the early nineteenth century when the highways were improved and robbers were less numerous, that walking became the pleasure and pastime of all classes' (*Travellers in Eighteenth-Century England* [London: John Murray, 1951] p. 22).
33. Capt. Robert Mignan, *Travels in Chaldaea, including a Journey from Bussorah to Baghdad, Hillah, and Babylon, Performed on Foot in 1827* (London: Henry Colburn & Richard Bentley, 1829); Capt. John Dundas Cochrane, *Narrative of a Pedestrian Journey through Russia and Siberian Tartary* (London: John Murray, 1824).
34. Anon., *A Tour on the Banks of the Thames from London to Oxford, in the Autumn of 1829. By a Pedestrian* (London, 1834) p. 2.
35. John Towner, 'The Grand Tour: A Key Phase in the History of Tourism', *Annals of Tourism Research*, XII (1985) 300. Towner is here summarising different opinions on the origins of the Tour.
36. Thomas Nugent, *The Grand Tour; or, a Journey through the Netherlands, Germany, Italy, and France*, 3rd ed., 4 vols. (London: J. Rivington et al., 1778) vol I, p. xi.
37. Jeremy Black, *The British Abroad: The Grand Tour in the Eighteenth Century* (Stroud: Alan Sutton, 1992) p. 168. See also R.S. Lambert, *Grand Tour: A Journey in the Tracks of the Age of Aristocracy* (London: Faber, 1935), and Christopher Hibbert, *The Grand Tour* (London: Weidenfeld, 1969).
38. Black, *British Abroad*, p. 300.
39. Towner, 'Grand Tour', 311, 316.
40. Thomas West, *A Guide to the Lakes in Cumberland, Westmorland and Lancashire*, 3rd ed. (1784; rpt Oxford: Woodstock Books, 1989) p. 1.
41. Wallace, *Walking*, p. 62.
42. Wallace, *Walking*, p. 65.
43. H.J. Dyos and D.H. Aldcroft, *British Transport: An Economic Survey from the Seventeenth Century to the Twentieth* (Harmondsworth: Penguin, 1974) p. 35. In the period cited, Dyos and Aldcroft say that 1100 separate trusts were created, dealing with 22 000 miles of road.
44. West, *Guide*, p. 2; W. Hutton, *Remarks upon North Wales, Being the Result of Sixteen Tours through that Part of the Principality* (Birmingham: Knott & Lloyd, 1803) pp. vii–viii.
45. Dyos and Aldcroft, *British Transport*, p. 75.
46. Philip S. Bagwell, *The Transport Revolution from 1770* (London: Batsford, 1974) p. 43.

Notes

47. Bagwell, *Transport*, p. 49; John Copeland, *Roads and Their Traffic, 1750–1850* (Newton Abbot: David & Charles, 1968) p. 85.
48. Bagwell, *Transport*, p. 48.
49. Dyos and Aldcroft, *British Transport*, p. 81.
50. John Thelwall, 'A Pedestrian Excursion through Several Parts of England and Wales during the Summer of 1797', *Monthly Magazine*, VIII (1799) 784.
51. Thelwall, 'Excursion', 785.
52. Leigh Hunt, 'Coaches', *Selected Essays*, ed. J.B. Priestley (London: Dent, 1929) pp. 126–7.
53. Eric J. Leed, *The Mind of the Traveller: From Gilgamesh to Global Tourism* (New York: Basic Books, 1991) p. 2.
54. Leed, *Mind*, p. 13.
55. Wallace, *Walking*, p. 29.
56. Charles P. Moritz, *Travels, Chiefly on Foot, through Several Parts of England, in 1782*, trans. by 'a Lady' (London: G.G. & J. Robinson, 1795) p. 122.
57. Moritz, *Travels*, pp. 173–4.
58. Moritz, *Travels*, p. 184.
59. Hucks, *Tour*, p. 5.
60. Thomas De Quincey, *Confessions of an English Opium Eater* [1856 revision] (Harmondsworth: Penguin, 1971) p. 185.
61. Warner, *Walk*, p. 7.
62. Thelwall, 'Excursion', 783.
63. John Bristed, $Aνθρωπλανομεος$; or *A Pedestrian Tour through Part of the Highlands of Scotlands, 1801*, 2 vols. (London: J. Wallis, 1803) vol I, p. i.
64. P. Stansbury, *A Pedestrian Tour of Two Thousand Three Hundred Miles, in North America* (New York: J.D. Myers & W. Smith, 1822) p. x (my italics).
65. Leed, *Mind*, pp. 13–14.

NOTES TO CHAPTER 2

1. Walker, *Remarks*, Advertisement; Hucks, *Tour*, p. 59; Penhouet, *Letters*, p. 41.
2. Bingley, *Tour*, vol. I, p. iv.
3. Anon., *Recollections of a Pedestrian*, 3 vols. (London: Saunders & Otley, 1826) vol. I, p. 13.
4. The road from London to Holyhead 'assumed a crucial political importance' after the Act of Union of 1801, and the improved road, centrally planned and funded from national taxes, was 'the nearest thing to national highway ever built'. Similarly, in Scotland 'the Jacobite risings in 1715 and 1745 had brought a reaction in the investment of public money in military roads and bridges to forestall further rebellion', and the new Commission for the Highland Roads established in 1803 built on these improvements (Dyos and Aldcroft, *British Transport*, pp. 82–3).

5. John Thelwall, *The Peripatetic; or, Sketches of the Heart, Of Nature and Society*, 3 vols. (London: published by the author, 1793) vol I, pp. 8–9.
6. Jean-Jacques Rousseau, *Reveries of the Solitary Walker* (1782; Harmondsworth: Penguin, 1979) p. 35.
7. Thelwall, *Peripatetic*, vol. I, p. 105.
8. Wallace, *Walking*, p. 61. See also pp. 92–5.
9. Leed, *Mind*, p. 72.
10. Jeffrey Robinson, *The Walk: Notes on a Romantic Image* (Norman, OK and London: University of Oklahoma Press, 1989) p. 52.
11. See the *Dictionary of National Biography*, vol. XIII, pp. 239–40.
12. *CLSTC*, vol. I, pp. 84, 90.
13. Hucks, *Tour* pp. 9, 27, 63.
14. See the seminal account in E.P. Thompson, *The Making of the English Working Class*, revised ed. (Harmondsworth: Penguin, 1980) chs. 1–5, esp. pp. 171–6.
15. Thelwall, 'Excursion', VIII, 532.
16. Thelwall, 'Excursion', IX, 229.
17. Thelwall, 'Excursion', VIII, 618.
18. Leed, *Mind*, p. 22.
19. Victor W. Turner, *The Ritual Process: Structure and Anti-Structure* (London: Routledge, 1969) pp. 94–5.
20. Turner, *Ritual*, pp. 95–6.
21. Hucks, *Tour*, p. 5.
22. Leed, *Mind*, p. 147.
23. Robert Southey, *Letters from England*, ed. Jack Simmons (London: Cresset Press, 1951) pp. 415–27.
24. *EY*, p. 168.
25. Leed, *Mind*, p. 11.
26. John Urry, *The Tourist Gaze: Leisure and Travel in Contemporary Societies* (London; Sage, 1990) p. 10.
27. The Claude glass was a hand-held plano-convex mirror that composed the scene into a picture. Thomas West recommended it to the visitor to the Lakes, saying that 'Where the objects are great and near, it removes them to a due distance, and shews them in the soft colours of nature'. Using one required the tourist to stand with his/her back to the view: 'It should be suspended by the upper part of the case, and the landscape will then be seen in the glass, by holding it a little to the right or left (as the position of the parts to be viewed require) and the face screened from the sun' (*Guide*, p. 12).
28. Erik Cohen, 'A Phenomenology of Tourist Experiences', *Sociology*, XIII (1979) 182.
29. Cohen, 'Phenomenology', 183, 185.
30. Cohen, 'Phenomenology', 188.
31. Hucks, *Tour*, pp. 61, 59.
32. I do not wish to overlook the fact that Wordsworth's description of the proud remains of 'primaeval' man in the Alps draws as much on his reading in travel books, including Coxe and Ramond, as it does on first-hand observation. Charles Norton Coe's proposal of a happy synthesis of experience and prior reading in *Descriptive Sketches* still

sounds measured and persuasive (*Wordsworth and the Literature of Travel* [New York: Bookman Associates, 1953] ch. 2).
33. Bingley, *Tour* vol. I, pp. 218–19.
34. Walker, *Remarks*, p. 70.
35. Thelwall, *Peripatetic*, vol. II, pp. 105–6.
36. Leed, *Mind*, pp. 182, 60.
37. I should note here that I am using the term 'philosophical travel' not in the same sense as Leed (for which see *Mind*, pp. 133–57), but more as equivalent to what he calls 'scientific travel'. My usage, however, parallels that in Romantic writing, for example the anonymous reviewer in the *Monthly Magazine* who praised William Coxe, on the publication of his *Travels in Switzerland*, as a 'philosophical traveller, who to the praise of accuracy joins the merit of science' (*Monthly Review*, Enlarged Series, I [1790] 32).
38. Leed, *Mind*, p. 73.
39. See Roman Jakobson, 'Two Aspects of Language and Two Types of Aphasic Disturbances', in Jakobson and Morris Halle, *Fundamentals of Language* (The Hague: Mouton, 1956).
40. Charles Burlington, David Llewellyn Rees and Alexander Murray, *The Modern Universal British Traveller* (London: J. Cooke, 1779) p. iii.
41. See Charles J. Batten, Jr., *Pleasurable Instruction: Form and Convention in Eighteenth-Century Travel Literature* (Berkeley and Los Angeles, CA: University of California Press, 1978) *passim*.
42. Robinson, *The Walk*, p. 67.
43. Hutton, *Remarks*, pp. 150–1.
44. Walker, *Remarks*, pp. 91–2.
45. Gray, 'Journal', printed as appendix to West, *Guide*, p. 201.
46. Ian Ousby comments on some of the parallels in his *The Englishman's England: Taste, Travel and the Rise of Tourism* (Cambridge: Cambridge University Press, 1990) pp. 146–8.
47. See West, *Guide*, pp. 98, 203–4.
48. See James T. Buzard, *The Beaten Track: European Tourism, Literature, and the Ways to 'Culture', 1800–1918* (Oxford: Clarendon Press, 1993) pp. 172–92. The fourth component of the authenticity effect, as analysed by Buzard, is 'picturesqueness', a term he uses in a less specialised sense than that which was current in the 1780s and 1790s, and which I shall examine in the next section.
49. Buzard, *Beaten Track*, p. 177.
50. Stanley Fish, *Doing What Comes Naturally: Change, Rhetoric, and the Practice of Theory in Literary and Legal Studies* (Oxford: Clarendon Press, 1989) p. 480.
51. Fish, *Doing*, p. 483.
52. Ann Bermingham, *Landscape and Ideology: The English Rustic Tradition, 1740–1860* (Berkeley and London: University of California Press, 1986) p. 70; Alan Liu, *Wordsworth: The Sense of History* (Stanford, CA: Stanford University Press, 1989) p. 94.
53. Liu, for example, when he comes to consider the Price-Repton-Knight dispute, finds the picturesque marking a 'leftward' drift in the politics of landscape 'toward an idea of revolution cognate with the American

or very early French Revolution' (Liu, *Wordsworth*, p. 113). A more single-mindedly libertarian interpretation of Knight's position is offered by Andrew Ballantyne, 'Turbulence and Repression: Re-reading *The Landscape*', in *The Picturesque Landscape* (Nottingham: Department of Geography, University of Nottingham, 1994) pp. 66–78.
54. Gilpin's unwavering way of referring to his tours and the written accounts of them, for example in *Observations, Relative Chiefly to Picturesque Beauty, Made in the Year 1772, on...the Mountains, and Lakes of Cumberland, and Westmoreland* (London: R. Blamire, 1786) p. xxii.
55. Kim Ian Michasiw, 'Nine Revisionist Theses on the Picturesque', *Representations*, 38 (Spring 1992) 95.
56. Christopher Hussey, *The Picturesque: Studies in a Point of View* (London: G.P. Putnam, 1927; rpt. London: Frank Cass, 1967) pp. 130–1.
57. See William Gilpin, *Observations on the River Wye* (London: R. Blamire, 1782) p. 7.
58. See William Gilpin, *Three Essays: On Picturesque Beauty; on Picturesque Travel; and on Sketching Landscape* (London: R. Blamire, 1792) pp. 16–18.
59. West, *Guide*, p. 105.
60. Richard Payne Knight, *The Landscape; a Didactic Poem* (London: W. Bulmer, 1794) I.317–28.
61. Gilpin, *Three Essays* p. 19.
62. Uvedale Price, *An Essay on the Picturesque, as compared with the Sublime and the Beautiful* (London: J. Robson, 1794) pp. 44–5, 18.
63. Edmund Burke, *A Philosophical Enquiry into the Origin of our Ideas of the Sublime and Beautiful* (Oxford: Oxford University Press, 1990) p. 140.
64. Burke, *Enquiry*, p. 105.
65. Burke, *Enquiry*, p. 141.
66. Gilpin, *Three Essays* p. 48.
67. Price, *Essay*, pp. 86–7.
68. I am alluding here to Knight, *The Landscape*, III.9–12.

NOTES TO CHAPTER 3

1. Hussey, *Picturesque*, p. 114.
2. Gilpin, *Observations on the River Wye*, p. 33.
3. The references are to *Observations on the River Wye*, pp. 35–6, and *Observations on Cumberland and Westmoreland*, vol. II, p. 44.
4. Michasiw, 'Nine Revisionist Theses', 94.
5. Gilpin, *Observations on Cumberland and Westmoreland*, pp. xxviii–xxix.
6. Gilpin, *Three Essays*, pp. 51–2.
7. Gilpin, *Three Essays*, p. 54.
8. Wallace, *Walking*, p. 49.
9. Wallace, *Walking*, pp. 43, 161.
10. Donald Appleyard, Kevin Lynch and John Myer, *The View from the Road* (Cambridge, MA: MIT Press, 1964) p. 4.

11. See Leed, *Mind*, pp. 74–80.
12. Interview with Richard Cork, in *Richard Long: Walking in Circles* (London: Thames & Hudson, 1991) pp. 251, 249.
13. Quoted in Dervla Murphy, *Where the Indus is Young: Walking in Baltistan* (London: John Murray, 1977) p. 141.
14. Bruce Chatwin, *The Songlines* (London: Jonathan Cape, 1987) pp. 62, 64.
15. See George Santayana, *The Sense of Beauty: Being the Outlines of Aesthetic Theory*, ed. William G. Holzberger and Herman J. Saatkamp (Cambridge, MA: MIT Press, 1988).
16. I quote from the text in Geoffrey Tillotson, Paul Fussell and Marshall Waingrow, ed., *Eighteenth-Century English Literature* (New York: Harcourt, Brace & World, 1969) pp. 785–9.
17. My text is from Tillotson, Fussell and Waingrow, *Eighteenth-Century English Literature*, pp. 809–10.
18. John Barrell, *The Idea of Landscape and the Sense of Place, 1730–1840: An Approach to the Poetry of John Clare* (Cambridge: Cambridge University Press, 1972) p. 36.
19. I quote from Alexander Chalmers, ed., *The Works of the English Poets*, 21 vols (London, 1810) vol. XIII, pp. 250–1. The poem is not line-numbered.
20. James Thomson, *The Seasons*, ed. James Sambrook (Oxford: Clarendon Press, 1981) *Spring*, l. 77.
21. William Crowe, *Lewesdon Hill*, 3rd ed. (London: T. Cadell & W. Davies, 1804) pp. 1–2. The poem is not line-numbered in this edition; subsequent page references will be included parenthetically in the main text.
22. *Collected Works of Oliver Goldsmith*, ed. Arthur Friedman, 5 vols. (Oxford: Clarendon Press, 1966) vol. IV, p. 247. Line references for *The Traveller* are to this edition and are cited in the body of the text.
23. Wallace, *Walking*, p. 91.
24. William Cowper, *The Task*, in *The Poetical Works of William Cowper*, ed. John Bruce, Aldine Poets edition, 3 vols (London: Bell & Daldy, 1866) I.135–6. Subsequent references (Book and line numbers) will be incorporated in the text.
25. Miss M. Bowen, 'The Walk', *Original Poems on Various Subjects* (Chepstow: S. Rogers, 1808). Line references are included parenthetically in the text.
26. A.R. Chisholm, 'La Pythie and its Place in Valéry's Work', *Modern Language Review* LVIII (1963) 23.
27. David Abercrombie, 'A Phonetician's View of Verse Structure', in *Studies in Phonetics and Linguistics* (London: Oxford University Press, 1965) p. 19.
28. D.W. Harding, *Words into Rhythm: English Speech Rhythm in Verse and Prose* (Cambridge: Cambridge University Press, 1976) p. 96.
29. Harding, *Words*, p. 101.
30. Harding, *Words*, p. 153.
31. Derek Attridge, *The Rhythms of English Poetry* (London and New York: Longman, 1982) pp. 60, 71, 73.

32. One does not, of course, make 'stressed' or 'unstressed' steps, but it is a common experience to *perceive* the sound of one's feet in walking as an alternation of strong and weak, just as one subjectively differentiates the objectively equal tick-tick-tick-tick of a clock into an alternating tick-tock-tick-tock.

NOTES TO CHAPTER 4

1. Thomas De Quincey, *Recollections of the Lakes and the Lake Poets*, ed. David Wright (Harmondsworth: Penguin, 1970) p. 135.
2. Donald E. Hayden, *Wordsworth's Walking Tour of 1790* (Tulsa, OK: University of Tulsa Press, 1983) p. 119.
3. See Wallace, *Walking*, ch. 3.
4. Seamus Heaney, 'The Makings of a Music: Reflections on Wordsworth and Yeats', in *Preoccupations: Selected Prose 1968–78* (London: Faber, 1980) p. 68.
5. I take my text from *William Wordsworth*, ed. Stephen Gill, The Oxford Authors (Oxford: Oxford University Press, 1984) pp. 220–3. Line references are included parenthetically in the main text.
6. William Hazlitt, *Selected Writings*, ed. Ronald Blythe (Harmondsworth: Penguin, 1970) p. 60.
7. My argument thus has points of similarity with the approach taken by Andrew Bennett in a fine essay, ' "Devious Feet": Wordsworth and the Scandal of Narrative Form', *ELH*, LIX (1992) 145–73, which advances 'the bafflement of linear pedestrian progression' (150) as the very 'generative force' (163) of Wordsworth's narrative verse. However, I disagree with Bennett that Wordsworth has 'nothing to say about walking *qua* walking' (164), as I shall make clear; conversely, I believe he overlooks a whole dimension of non-scandalous arrested motion in Wordsworth's poetry, one that focuses the ambivalent desire for stability, emplacement and belonging.
8. Robert A. Aubin, *Topographical Poetry in Eighteenth-Century England* (New York: Modern Language Association, 1936) p. 219.
9. *The Poetical Works of William Wordsworth*, ed. E. de Selincourt and Helen Darbishire, 5 vols (Oxford: Clarendon Press, 1940–9) vol. I, p. 319.
10. J.R. Watson, *Picturesque Landscape and English Romantic Poetry* (London: Hutchinson, 1970) p. 67.
11. All references to *An Evening Walk* are to the reading text of the 1793 edition presented in the Cornell Wordsworth volume edited by James Averill (Ithaca, NY: Cornell University Press, 1983). Line references are included parenthetically in the main text.
12. Liu, *Wordsworth*, p. 132. More generally, Liu argues that eighteenth-century tours typically appear undermotivated owing to their reliance on such 'point-scenes' (pp. 5–6).
13. Kim Taplin, *The English Path* (Woodbridge: The Boydell Press, 1979) p. 140.

14. Gerald Izenberg has noted this peculiarity. See his fine discussion in *Impossible Individuality: Romanticism, Revolution, and the Origins of Modern Selfhood, 1787–1802* (Princeton, NJ: Princeton University Press, 1992) pp. 156–61.
15. *EY*, p. 32.
16. All references to *Descriptive Sketches* are to the reading text of the 1793 edition provided in the Cornell Wordsworth volume edited by Eric Birdsall (Ithaca, NY: Cornell University Press, 1984). Line references are included parenthetically in the main text.
17. *EY*, p. 36.
18. *EY*, pp. 35–6.
19. Izenberg, *Impossible Individuality*, p. 175.
20. Izenberg, *Impossible Individuality*, p. 175.
21. *EY*, p. 109.
22. All references to the Salisbury Plain poems are to the reading texts presented in the Cornell Wordsworth volume edited by Stephen Gill (Ithaca, NY: Cornell University Press, 1975). Line references are included parenthetically in the main text.
23. *The Prelude* (1805), ed. E. de Selincourt, corrected Stephen Gill (London: Oxford University Press, 1970) XII.323. All references to *The Prelude* are to this edition, and from now on will be included parenthetically in the main text.
24. Theresa M. Kelley, *Wordsworth's Revisionary Aesthetics* (Cambridge: Cambridge University Press, 1988) p. 125.
25. David Simpson, *Wordsworth's Historical Imagination: The Poetry of Displacement* (London: Methuen, 1987) p. 44.
26. I quote from the early version of 1796–97 presented in *The Borderers*, ed. Robert Osborn, The Cornell Wordsworth (Ithaca, NY: Cornell University Press, 1982).
27. All references to *The Ruined Cottage* are to the reading text of the MS. B version of 1798 presented in *'The Ruined Cottage' and 'The Pedlar'*, ed. James Butler, The Cornell Wordsworth (Ithaca, NY: Cornell University Press, 1979). Line references are included parenthetically in the main text.
28. *Observations on Cumberland and Westmoreland*, vol. II, p. 44.
29. All references to *Home at Grasmere* are to the early MS. B version presented in the Cornell Wordsworth volume edited by Beth Darlington (Ithaca, NY: Cornell University Press, 1977). Line references are included parenthetically in the main text.
30. Wallace, *Walking*, p. 120.
31. Price, *Essay*, p. 98.
32. John Elder, *Imagining the Earth: Poetry and the Vision of Nature* (Urbana and Chicago, IL: University of Illinois Press, 1985) pp. 93–4.
33. Wallace, *Walking*, p. 120. See also Bennett, '"Devious Feet"', 166–8, for a reading of Wordsworth's 'anxiety about wandering' as rooted in a social 'anxiety of vagrancy'.
34. Kenneth R. Johnston, *Wordsworth and 'The Recluse'* (New Haven, CT and London: Yale University Press, 1984) pp. 81–3.
35. See Simpson, *Wordsworth's Historical Imagination*, pp. 115–21.

36. Elder, *Imagining* p. 98.
37. Elder, *Imagining*, p. 99.
38. M.H. Abrams, *Natural Supernaturalism: Tradition and Revolution in Romantic Literature* (New York: Norton, 1971) p. 284. On the peripatetic structure of *The Prelude*, see also Herbert Lindenberger, *On Wordsworth's 'Prelude'* (Princeton, NJ: Princeton University Press, 1963) pp. 305–7.
39. Leed, *Mind*, p. 56.
40. Johnston, *Wordsworth*, pp. 156–7.
41. Johnston, *Wordsworth*, pp. 149–58; Mark Reed, 'The Speaker of *The Prelude*', in *Bicentenary Wordsworth Studies in Memory of John Alban Finch*, ed. Jonathan Wordsworth (Ithaca, NY: Cornell University Press, 1970) pp. 283–9.
42. Johnston, *Wordsworth*, p. 150.
43. Elder, *Imagining*, p. 98.
44. Geoffrey H. Hartman, *Wordsworth's Poetry 1787–1814* (New Haven, CT and London: Yale University Press, 1971) p. 17.
45. Hartman, *Wordsworth's Poetry*, p. 47.
46. Hartman, *Wordsworth's Poetry*, p. xvi.

NOTES TO CHAPTER 5

1. *CLSTC*, vol. I, pp. 259–60; Hazlitt, *Selected Writings*, pp. 52– 60; *Letters of William and Dorothy Wordsworth: The Later Years, Part IV, 1840–1853*, ed. E. de Selincourt, 2nd ed., revised Alan G. Hill (Oxford: Clarendon Press, 1988) p. 719; *CLSTC*, vol. I, p. 84.
2. See Joseph Cottle, *Early Recollections; Chiefly Relating to the Late Samuel Taylor Coleridge*, 2 vols. (London: Longman, Rees & Co., 1837) vol. I, pp. 40–51.
3. The latter expedition has prompted a work in the 'footsteps' genre of travel writing. See Alan Hankinson, *Coleridge Walks the Fells: A Lakeland Journey Retraced* (Maryport, Cumbria: Ellenbank Press, 1991).
4. *CLSTC*, vol. II, pp. 978, 982. Coleridge was afflicted nightly on this tour by nightmares caused by opium withdrawal, and wrote of 'substantial Misery foot-thick'. See Richard Holmes, *Coleridge: Early Visions* (London: Hodder & Stoughton, 1989) pp. 352–5.
5. *CLSTC*, vol. I, pp. 89, 91; Hucks, *Tour* pp. 14–15, lxi.
6. Hucks, *Tour* pp. 7, 61.
7. Hucks, *Tour* pp. 42–3.
8. *CLSTC*, vol. I, p. 88. Nicholas Roe usefully retells the story of the origins and collapse of the Pantisocracy scheme from Southey's point of view in *The Politics of Nature: Wordsworth and Some Contemporaries* (Basingstoke and London: Macmillan, 1992) pp. 36–55.
9. *CLSTC*, vol. I, pp. 84, 88.
10. *CLSTC*, vol. I, pp. 83, 88.
11. Burke, *Enquiry*, p. 73.

12. *CLSTC*, vol. I, p. 84. This poem was included in Coleridge's letter to Southey of 6 July 1794 but never published in his lifetime.
13. *CLSTC*, vol. I, p. 83.
14. See Patricia Ball, *The Science of Aspects: The Changing Role of Fact in the Work of Coleridge, Ruskin and Hopkins* (London: Athlone Press, 1971); Raimondo Modiano, *Coleridge and the Concept of Nature* (London and Basingstoke: Macmillan, 1985), esp. ch. 1; William Ruddick, '"As much diversity as the heart that trembles": Coleridge's Notes on the Lakeland Fells', in *Coleridge's Imagination: Essays in Memory of Pete Laver*, ed. Richard Gravil, Lucy Newlyn and Nicholas Roe (Cambridge: Cambridge University Press, 1985) pp. 88–101; and Harold D. Baker, 'Landscape as Textual Practice in Coleridge's *Notebooks*', *ELH*, LIX (1992) 651– 70.
15. Baker, 'Landscape', 667.
16. Modiano, *Coleridge* p. 13; Ruddick, 'As much diversity', 91.
17. See, for example, *The Notebooks of Samuel Taylor Coleridge*, ed. Kathleen Coburn, 4 double vols to date (London: Routledge & Kegan Paul, 1957–) vol. I, 1215. Henceforth cited as *NSTC*; all references to the *Notebooks* are to volume and *entry number*.
18. See Modiano, *Coleridge*, p. 13.
19. *NSTC*, vol. I, 1212.
20. See, for example, the sketch-map of Wasdale in *NSTC* vol. I, 1213.
21. *CLSTC*, vol. I, p. 513.
22. For Freud's somewhat cryptic comments on the ego as 'a mental projection of the surface of the body', see *The Ego and the Id*, in *On Metapsychology: The Theory of Psychoanalysis*, trans. James Strachey, ed. Angela Richards, Pelican Freud Library, vol. XI (Harmondsworth: Penguin, 1984) pp. 364–5.
23. See Ball, *Science*, p. 22.
24. See Michael G. Cooke, 'The Manipulation of Space in Coleridge's Poetry', in *New Perspectives in Coleridge and Wordsworth*, ed. Geoffrey Hartman (New York: Columbia University Press, 1972) pp. 165–94.
25. *NSTC*, vol. I, 1771.
26. *NSTC*, vol. I, 556.
27. Baker, 'Landscape', 657.
28. Baker, 'Landscape', 667.
29. *CLSTC*, vol. II, p. 918.
30. *CLSTC*, vol. II, p. 916.
31. *NSTC*, vol. I, 537.
32. Donald Appleyard, Kevin Lynch and John R Myer, *The View from the Road* (Cambridge, MA: MIT Press, 1964) pp. 17, 6.
33. *NSTC*, vol. I, 1213.
34. *NSTC*, vol. I, 1610.
35. Modiano, *Coleridge*, p. 38.
36. The phrase is Richard Holmes's. See Holmes, *Coleridge*, p. 281.
37. All references to Coleridge's poems, unless otherwise indicated, are to the one-volume *Poetical Works*, ed. E.H. Coleridge (London: Oxford University Press, 1967). Line references are included parenthetically in the main text.

38. Jay Appleton, *The Experience of Landscape* (Chichester: John Wiley, 1975) p. 69. Anne K. Mellor adverts briefly to Appleton's work in 'Coleridge's "This Lime-Tree Bower My Prison" and the Categories of English Landscape', *Studies in Romanticism*, XVIII (1979) 265. For an attempt to re-synthesise Appleton's insights within a broader hermeneutic model encompassing 'cultural rules' and 'personal strategies' in addition to biological laws, see Steven C. Bourassa, *The Aesthetics of Landscape* (London and New York: Belhaven Press, 1991).
39. Appleton, *Experience*, p. 170.
40. Cooke, 'Manipulation of Space', 176.
41. Bourassa, *Aesthetics*, p. 92.
42. Cooke, 'Manipulation of Space', 183.
43. Kelvin Everest, *Coleridge's Secret Ministry: The Context of the Conversation Poems 1795–1798* (Hassocks, Sussex: Harvester Press, 1979) p. 237.
44. Everest, *Coleridge's Secret Ministry*, p. 251; Modiano, *Coleridge*, pp. 86–7; Mellor, 'Coleridge's "This Lime-Tree Bower"', *passim*.
45. Mellor, 'Coleridge's "This Lime-Tree Bower"', 268.
46. *CLSTC*, vol. I, p. 504. I quote the 'Lines' from the version Coleridge included in his letter to Sara of 17 May 1799.
47. James McKusick, 'From Coleridge to John Muir: The Romantic Origins of Environmentalism', *The Wordsworth Circle*, XXVI (1995) 38. McKusick uses this phrase of the philosophy of Muir; but, in focusing on the latter's use of the Aolian harp metaphor, argues for his deep indebtedness to Coleridge. In a similar vein, William Rossi has established a Coleridgean influence on Henry Thoreau's advocacy of a 'Transcendental ecology' ('"The Limits of an Afternoon Walk": Coleridgean Polarity in Thoreau's "Walking"', *ESQ*, XXXIII, ii [1987] 100).
48. Modiano, *Coleridge* p. 88.
49. The first published text of the poem uses the image of a 'deep embrace'.
50. See Modiano, *Coleridge*, pp. 89–93.
51. H.R. Rookmaaker, Jr., *Towards a Romantic Conception of Nature: Coleridge's Poetry up to 1803* (Amsterdam: John Benjamins, 1984) p. 163.
52. Coleridge uses the Brocken spectre as a metaphor for the mind misrecognising its own constructions in 'Constancy to an Ideal Object', ll. 29–32: the woodman 'Sees full before him, gliding without tread, / An image with a glory round its head; / The enamoured rustic worships its fair hues, / Nor know he makes the shadow, he pursues!'

NOTES TO CHAPTER 6

1. Sarah Hazlitt, 'Journal of My Trip to Scotland', *University of Buffalo Studies*, XXIV (1959) 208.
2. 'Journal', 215.
3. 'Journal', 202–3, 204, 208.
4. 'Journal', 201.
5. 'Journal', 216.

6. 'Journal', 215. Shirley Foster, in her study of nineteenth-century women's travel writing, suggests that reference to sexual dangers would have been omitted from such writing in order to uphold the required standard of 'literary femininity' (*Across New Worlds: Nineteenth-Century Women Travellers and their Writings* [Hemel Hempstead: Harvester Wheatsheaf, 1990] p. 19), but there is less reason to posit such self-censorship in a text genuinely not intended for publication.
7. 'Journal', 220.
8. Mary Shelley, *History of a Six Weeks' Tour* (1817; rpt Oxford: Woodstock Books, 1989). This text narrates Mary's elopement journey with Percy Shelley (and Claire Clairemont) in 1814: it began as an attempt to walk through France into Switzerland, which she describes interestingly as an enterprise which in England they 'could not have put...in execution without sustaining continual insult and impertinence' (p. 13). However, as early as Troyes Percy Shelley suffers a sprained ankle which 'rendered our pedestrianism impossible' (p. 26), and they proceed therefrom in an 'open *voiture*'.
9. Jane Robinson, *Wayward Women: A Guide to Women Travellers* (Oxford: Oxford University Press, 1990).
10. James Clifford, 'Travelling Cultures', in *Cultural Studies*, ed. Lawrence Grossberg, Cary Nelson and Paula A. Treichler (London: Routledge, 1992) pp. 106, 105.
11. Foster, *Across New Worlds*, p. 11.
12. Leed, *Mind*, pp. 115–16.
13. See Black, *British Abroad*, pp. 199–200.
14. Roy Porter, '"All madness for writing": John Clare and the asylum', in *John Clare in Context*, ed. Hugh Haughton, Adam Phillips and Geoffrey Summerfield (Cambridge: Cambridge University Press, 1994) p. 263.
15. *John Clare By Himself*, ed. Eric Robinson and David Powell (Ashington, Northumberland and Manchester: MidNAG/Carcanet, 1996) p. 257. Unless indicated otherwise, all references to Clare's prose are to this edition, abbreviated *JCBH*.
16. *JCBH*, p. 258.
17. Ronald Blythe, '*Solvitur Ambulando*: John Clare and Footpath Walking', *The John Clare Society Journal*, XIV (July 1995) 18.
18. *JCBH*, pp. 260, 261.
19. *JCBH*, pp. 263, 339.
20. Leed, *Mind*, p. 28.
21. Barrell, *Idea of Landscape*, pp. 63, 92.
22. Clifford, 'Travelling Cultures', 108. The argument above is summarised from earlier sections of the same essay.
23. See Meena Alexander, 'Dorothy Wordsworth: The Grounds of Writing', *Women's Studies*, XIV (1988) 195–210; and *Women in Romanticism* (Basingstoke: Macmillan, 1989) esp. pp. 95–9, 172–8.
24. *EY*, pp. 25, 46–7, 114, 117.
25. *Journals of Dorothy Wordsworth*, 2nd ed., ed Mary Moorman (Oxford: Oxford University Press, 1971) pp. 11–12. All references to the *Alfoxden*

Journal and *Grasmere Journal* are to this edition, and from now on will be included parenthetically in the text.

26. Susan Levin, *Dorothy Wordsworth and Romanticism* (New Brunswick, NJ and London: Rutgers University Press, 1989) p. 21. Anita Hemphill McCormick's '"I shall be beloved – I want no more": Dorothy Wordsworth's Rhetoric and the Appeal to Feeling in *The Grasmere Journals*', *Philological Quarterly*, LXIX (1990) 471– 93, draws out the implications of Wordsworth's position as the implied reader of the journal.
27. Alexander, 'Dorothy Wordsworth', 205–6, comments sadly on the compulsive fidelity with which, in her late poem, 'Thoughts on my sick-bed', Dorothy Wordsworth fulfils the script that her brother had written for her in 'Tintern Abbey', and finds her own voice 'held *within* the space created by the poem'.
28. ll. 9–12. I am quoting from the text of this poem provided in Levin, *Dorothy Wordsworth*, pp. 184–7. The Appendix to Levin's study constitutes the first collected edition of Dorothy Wordsworth's poems.
29. *Home at Grasmere*, ll. 848, 858; p. 92.
30. Levin, *Dorothy Wordsworth*, p. 38.
31. Anne K. Mellor, *Romanticism and Gender* (London: Routledge, 1993) pp. 159, 160.
32. *Journal of a Tour on the Continent*, in *Journals of Dorothy Wordsworth*, ed. E. de Selincourt, 2 vols. (London: Macmillan, 1952) vol. II, p. 282. References to this edition (abbreviated *JTC* where necessary) will from now on be included in the text.
33. See, for example, Meena Alexander, 'Dorothy Wordsworth: The Grounds of Writing', *Women's Studies*, XIV (1988) 201–2.
34. Levin, *Dorothy Wordsworth*, pp. 97–9.
35. 'When shall I tread your garden path?', l. 5; Levin, *Dorothy Wordsworth*, p. 231.
36. Linda Mills Woolsey, 'Houseless Woman and Travelling Lass: Mobility in Dorothy Wordsworth's *Grasmere Journals*', *Tennessee Philological Bulletin*, XXVII (1990) 31.
37. McCormick, '"I shall be beloved"', 482.
38. Dorothy Wordsworth, *Recollections of a Tour Made in Scotland (1803)*, in *Journals of Dorothy Wordsworth*, ed. E. de Selincourt, vol. I, p. 314. References to the *Recollections*, abbreviated *RTS* where necessary, will from now on be included in the text.
39. Alexander, *Women in Romanticism*, p. 109.
40. John Nabholtz, 'Dorothy Wordsworth and the Picturesque', *Studies in Romanticism*, III (1964) 122–4.
41. See Levin, *Dorothy Wordsworth*, pp. 104–8.
42. Alexander, 'Dorothy Wordsworth', 196–7, 198.
43. See *The Wordsworth Circle*, XXVI (1995) 231.
44. ll. 25–6. Quotations from all Clare's poems are taken from the Oxford Authors *John Clare*, ed. Eric Robinson and David Powell (Oxford: Oxford University Press, 1984). Line references will from now on be provided parenthetically in the main text.
45. *JCBH*, p. 134.
46. Blythe, '*Solvitur Ambulando*', 21.

47. *JCBH*, p. 78.
48. *JCBH*, pp. 83–4.
49. James McKusick, 'Beyond the Visionary Company: John Clare's Resistance to Romanticism', in *John Clare in Context*, ed. Haughton *et al.*, pp. 232–4.
50. See *JCBH*, pp. 90–2.
51. Bob Bushaway, *By Rite: Custom, Ceremony and Community in England 1700–1800* (London: Junction Books, 1982) p. 82. For more details, see also A.R. Wright, *British Calendar Customs: England*, 3 vols (London: William Glaisher, 1936–40) vol I, pp. 129–37.
52. George Deacon (ed.), *John Clare and the Folk Tradition* (London: Sinclair Browne, 1983) p. 285. I am indebted to John Goodridge for this reference.
53. Bushaway, *By Rite*, p. 84; for Barrell's masterly analysis of the circular idea of space peculiar to the unimproved landscape, see *Idea of Landscape*, ch. 3.
54. John Goodridge and Kelsey Thornton, 'John Clare: The Trespasser', in *John Clare in Context*, ed. Haughton *et al.*, pp. 103, 99.
55. Barrell, *Idea of Landscape*, p. 87.
56. Seamus Heaney, 'John Clare: A Bi-centenary Lecture', in *John Clare in Context*, ed. Haughton *et al.*, p. 137.
57. See Mark Storey's summary of Taylor's advice to Clare in 'Clare and the Critics', in *John Clare in Context*, ed. Haughton *et al.*, pp. 42–3; Gilbert, *Walks*, p. 35; Barrell, *Idea of Landscape*, p. 152; Timothy Brownlow, *John Clare and Picturesque Landscape* (Oxford: Clarendon Press, 1983) pp. 81–2, 95.
58. Barrell, *Idea of Landscape*, pp. 162, 171, 157.
59. See Jonathan Bate, *Romantic Ecology: Wordsworth and the Environmental Tradition* (London: Routledge, 1991), and 'The Rights of Nature', *John Clare Society Journal*, XIV (July 1995) 7– 15.
60. Bushaway, *By Rite*, p. 99.
61. Lawrence Buell, *The Environmental Imagination: Thoreau, Nature Writing, and the Formation of American Culture* (Cambridge, MA: Harvard University Press, 1995) p. 267; but my commentary draws on different sections of the book.
62. Buell, *Environmental Imagination*, p. 218.
63. *JCBH*, pp. 40–1.

NOTES TO CHAPTER 7

1. *Selected Writings*, pp. 137–8. Subsequent references to this essay will be included parenthetically in the main text.
2. *Letters of John Keats*, ed. Robert Gittings (Oxford: Oxford University Press, 1970) p. 83. All references to the letters and poems written on Keats's tour are to this edition (henceforth cited as *LJK*), and from now on will be included parenthetically in the main text. The letters are reproduced, along with Brown's more conventional unfinished

account, in Carol Kyros Walker, *Walking North with Keats* (New Haven, CT and London: Yale University Press, 1992).
3. Stuart Sperry, *Keats the Poet* (Princeton, NJ: Princeton University Press, 1973) pp. 138, 153.
4. 'I put down Mountains, Rivers Lakes, dells, glens, Rocks, and Clouds, With beautiful enchanting, gothic picturesque fine, delightful, enchancting, Grand, sublime – a few Blisters etc – and now you have our journey thus far' (letter to J.H. Reynolds, 11, 13 July 1818, *LJK*, p. 120).
5. For more information on Brown, see Walker, *Walking*, pp. 5–8.
6. Robert Gittings, *John Keats* (1968; rpt Harmondsworth: Penguin, 1979) pp. 651–2.
7. Brown notes one interesting disagreement in the section of his journal dealing with the road between Dumfries and Kirkcudbright. He objects to the labouring-class women spoiling their 'neatness of attire' by not wearing shoes and stockings, but 'Keats was of an opposite opinion, and expiated on the beauty of a human foot, that had grown without an unnatural restraint' (Walker, *Walking*, p. 238).
8. In his letter to Benjamin Bailey of 18–22 July, Keats confesses that he has 'not a right feeling towards Women', and that he is 'happier alone among Crowds of men, by myself or with a friend or two' (*LJK*, p. 136).
9. McGann uses this term, in contradistinction to 'linguistic codes', to refer to all aspects of the material production and distribution of a text that have signifying potential. See *The Textual Condition* (Princeton, NJ: Princeton University Press, 1991) esp. ch. 3.
10. See Walker, *Walking*, pp. 150–1.
11. The author, 'Z', was in fact John Gibson Lockhart.
12. Jeffrey N. Cox, 'Keats in the Cockney School', *Romanticism*, II (1996) 28.
13. Cox, 'Keats', 29.
14. Cox, 'Keats', 33–4.
15. William Keach, 'Cockney Couplets: Keats and the Politics of Style', *Studies in Romanticism*, XXV (1986) 183, 189. Keach quotes from Hunt's *Examiner* review of Keats's 1817 *Poems*.
16. See Marilyn Butler, *Romantics, Rebels and Reactionaries* (Oxford: Oxford University Press, 1981) ch. 5, 'The Cult of the South'.
17. Walker, *Walking*, p. 208. Gittings's distaste for this composition is such that he omits it from his version in the *Letters*.
18. Mikhail Bakhtin, 'From the Prehistory of Novelistic Discourse', in *The Dialogic Imagination: Four Essays*, trans. Caryl Emerson and Michael Holquist (Austin, TX: University of Texas Press, 1981) p. 55.
19. Cox notes that the *Blackwood's* attacks referred to Hunt, in an allusion to popular ritual, as the 'King of Cockaigne', and saw Keats as a 'courtier' at Hunt's court ('Keats', 30).
20. 'Cockney School of Poetry No. IV', *Blackwood's Edinburgh Magazine*, August 1818, in *Romantic Bards and British Reviewers*, ed. J.O. Hayden (London: Routledge & Kegan Paul, 1971) p. 319.
21. Sperry, *Keats*, p. 138.
22. Walker, *Walking*, p. 24.

23. From the *Blackwood's* review cited in note 20 (Hayden, *Romantic Bards*, p. 321).
24. Keach, 'Cockney Couplets', 190.
25. Walker, *Walking*, p. 229.
26. The phrase is Jeffrey Robinson's. See *The Walk*, p. 83.
27. Raymond Williams, *The Country and the City* (London: Chatto & Windus, 1973) p. 233.
28. I have used the text in *The Poetical Works of John Gay*, ed. G.C. Faber (London: Oxford University Press, 1926).
29. See 'On Some Motifs in Baudelaire', in *Illuminations*, ed. Hannah Arendt, trans. Harry Zohn (London: Jonathan Cape, 1970) pp. 152–96.
30. Williams, *Country*, p. 150.
31. William Hazlitt, *Liber Amoris*, in *Selected Writings*, p. 369.
32. Williams, *Country*, p. 151.
33. De Quincey, *Confessions*, p. 45. Subsequent references to the *Confessions* will be included parenthetically in the main text.
34. Walter Benjamin, *Charles Baudelaire: A Lyric Poet in the Era of High Capitalism*, trans. Harry Zohn (London: New Left Books, 1973) p. 55.
35. Benjamin, *Illuminations*, p. 169.
36. *Selected Essays*, p. 78.
37. *Selected Essays*, p. 24.
38. 'A Ramble in Mary-le-Bone', in *Leigh Hunt's Political and Occasional Essays*, ed. Lawrence Huston Houtchens and Carolyn Washburn Houtchens (New York and London: Columbia University Press, 1962) pp. 279–96; *A Saunter in the West End* (London: Hurst & Blackett, 1861) p. 2.
39. 'Of the Sight of Shops', *Selected Essays*, pp. 25–6.
40. *Selected Essays*, p. 60–1.
41. *Selected Essays*, p. 16.
42. *Selected Essays*, p. 36.
43. *Selected Essays*, p.229. Further page references for this short essay are unnecessary.
44. Robinson, *The Walk*, p. 83.
45. *The Poetical Works of Leigh Hunt*, ed. H. S. Milford (Oxford: Oxford University Press, 1923) p. 269. Line references to the poem will be included parenthetically in the main text.
46. The phrase is the *London Magazine*'s; cited in Cox, 'Keats', 33.

Bibliography

PRIMARY SOURCES

A.A. 'A Pedestrian Tour in North Wales.' *Monthly Magazine*, I (1796) 15–18, 104–7, 191–4.
Anon. *The Life and Adventures of the Celebrated Walking Stewart*. London: E. Wheatley, 1822.
——. *The Observant Pedestrian; Or, Traits of the Heart*. London: William Lane, 1795.
——. *Recollections of a Pedestrian*. London: Saunders and Otley, 1826.
——. *A Short Sketch of the Life of Mr Foster Powell, The Great Pedestrian*. London: R.H. Westley, 1793.
——. *A Tour on the Banks of the Thames from London to Oxford*. London: Printed for the Author, 1834.
Bingley, Rev. W. *A Tour Round North Wales, Performed during the Summer of 1798*. 2 vols. London: E. Williams, 1800.
Bowen, Miss M. *Original Poems*. Chepstow: S. Rogers, 1808.
Bowles, William Lisle. *Sonnets, Written Chiefly on Picturesque Spots, During a Tour*. 2nd ed. Bath: R. Cruttwell, 1789.
Bristed, John. $Aνθρωπλανομεος$; *or A Pedestrian Tour through Part of the Highlands of Scotlands, 1801*. 2 vols. London: J. Wallis, 1803.
Budworth, Joseph. *A Fortnight's Ramble to the Lakes in Westmoreland, Lancashire, and Cumberland*. London: Hookham & Carpenter, 1792.
Burke, Edmund. *A Philosophical Enquiry into the Origin of our Ideas of the Sublime and Beautiful*. Ed. Adam Phillips. Oxford: Oxford University Press, 1990.
Burlington, Charles, David Llewellyn Rees and Alexander Murray. *The Modern Universal British Traveller*. London: Printed for J. Cooke, 1779.
Clare, John. *John Clare*. Ed. Eric Robinson and David Powell. Oxford Authors. Oxford: Oxford University Press, 1984.
——. *John Clare By Himself*. Ed. Eric Robinson and David Powell. Ashington, Northumberland and Manchester: MidNAG/Carcanet, 1996.
Cochrane, Capt. John. *Narrative of a Pedestrian Journey through Russia and Siberian Tartary*. London: John Murray, 1824.
Coleridge, Samuel T. *Collected Letters of Samuel Taylor Coleridge*. Ed. E.L. Griggs. 6 vols. Oxford: Clarendon Press, 1956–71.
——. *The Notebooks of Samuel Taylor Coleridge*. Ed. Kathleen Coburn. 4 double vols to date. London: Routledge & Kegan Paul, 1957– .
——. *Poetical Works*. Ed. E.H. Coleridge. London: Oxford University Press, 1967.

Cottle, Joseph. *Early Recollections; Chiefly Relating to the Late Samuel Taylor Coleridge*. 2 vols. London: Longman, Rees & Co., 1837.
Cowper, William. *Poetical Works*. Ed. John Bruce. Aldine Poets. 3 vols. London: Bell & Daldy, 1866.
Coxe, William. *Travels in Switzerland and in the Country of the Grisons*. 2 vols. Paris: James Decker, 1802.
Crowe, William. *Lewesdon Hill*. 3rd ed. London: T. Cadell & W. Davies, 1804.
De Quincey, Thomas. *Confessions of an English Opium Eater*. Ed. Alethea Hayter. Harmondsworth: Penguin, 1971.
——. *Recollections of the Lakes and the Lake Poets*. Ed. David Wright. Harmondsworth: Penguin, 1970.
——. *The Works of Thomas De Quincey*. 16 vols. Edinburgh: A. & C. Black, 1862–78.
Dyer, John. *Works of John Dyer*. In *The Works of the English Poets*. Ed. Alexander Chalmers. 21 vols. London, 1810. Vol. XIII.
Gay, John. *Poetical Works*. Ed. G.C. Faber. London: Oxford University Press, 1926.
Gilbert, Joseph and Thomas Howell. *A Correct and Minute Journal of the Time Occupied in Every Mile By Mr John Stokes, of Bristol, During his Walk of Fifty Miles per Day for Twenty Successive Days, making One Thousand Miles*. Bristol: C. Frost et al., 1815.
Gilpin, William. *Observations on the River Wye, and Several Parts of South Wales, Relative Chiefly to Picturesque Beauty; Made in the Summer of 1770*. London: R. Blamire, 1782.
——. *Observations, Relative Chiefly to Picturesque Beauty, Made in the Year 1772, on Several Parts of England; Particularly the Mountains, and Lakes of Cumberland, and Westmoreland*. London: R. Blamire, 1786.
——. *Three Essays: On Picturesque Beauty; on Picturesque Travel; and on Sketching Landscape*. London: R. Blamire, 1792.
Goldsmith, Oliver. *Collected Works*. Ed. Arthur Friedman. 5 vols. Oxford: Clarendon Press, 1966.
Hazlitt, Sarah. 'Journal of My Trip to Scotland.' *University of Buffalo Studies*, XXIV (1959) 185–252.
Hazlitt, William. *Selected Writings*. Ed. Ronald Blythe. Harmondsworth: Penguin, 1970.
Hucks, Joseph. *A Pedestrian Tour through North Wales*. London: J. Debrett & J. Edwards, 1795.
Hunt, Leigh. *Leigh Hunt's Political and Occasional Essays*. Ed. Lawrence Huston Houtchens and Carolyn Washburn Houtchens. New York: Columbia University Press, 1962.
——. *Poetical Works*. Ed. H.S. Milford. Oxford: Oxford University Press, 1923.
——. *A Saunter in the West End*. London: Hurst & Blackett, 1861.
——. *Selected Essays*. Ed. J.B. Priestley. London: Dent, 1929.
Hutton, William. *Remarks upon North Wales, Being the Result of Sixteen Tours through that Part of the Principality*. Birmingham: Knott & Lloyd, 1803.
Keats, John. *Letters of John Keats*. Ed. Robert Gittings. London: Oxford University Press, 1970.

Kelly, Michael. *Reminiscences of Michael Kelly, of the King's Theatre, and Theatre Royal Drury Lane, Including a Period of Nearly Half a Century*. 2nd ed. 2 vols. London: Henry Colburn, 1826.

Knight, Richard Payne. *The Landscape: A Didactic Poem*. London: W. Bulmer, 1794.

Mignan, Capt. Robert. *Travels in Chaldaea, including a Journey from Bussorah to Bagdad, Hillah, and Babylon, Performed on Foot in 1827*. London: H. Colburn & R. Bentley, 1829.

Moritz, Charles P. *Travels, Chiefly on Foot, through Several Parts of England, 1782, Described in Letters to a Friend*. London: G. G. & J. Robinson, 1795.

Newell, Rev. R.H. *Letters on the Scenery of Wales*. London: Baldwin, Cradock & Joy, 1821.

Nugent, Thomas. *The Grand Tour; or, a Journey through the Netherlands, Germany, Italy and France*. 3rd ed. 4 vols. London: J. Rivington et al., 1778.

Penhouet, A.B.L. Mandet de. *Letters Describing a Tour through Part of South Wales. By a Pedestrian Traveller*. London: J. Edwards et al., 1797.

Price, Uvedale. *An Essay on the Picturesque, as Compared with the Sublime and the Beautiful; and, on the Use of Studying Pictures, for the Purpose of Improving Real Landscape*. London: J. Robson, 1794.

Rousseau, Jean-Jacques. *Reveries of the Solitary Walker*. Trans. Peter France. Harmondsworth: Penguin, 1979.

Shelley, Mary. *History of a Six Weeks' Tour*. 1817. Repr. Oxford: Woodstock Books, 1991.

Skrine, Henry. *Three Successive Tours in the North of England, and Great Part of Scotland*. London: P. Elmsly, 1795.

Southey, Robert. *Letters from England*. Ed. Jack Simmons. London: Cresset Press, 1951.

Stansbury, P. *A Pedestrian Tour of Two Thousand Three Hundred Miles, in North America*. New York: J.D. Myers & W. Smith, 1822.

Stewart, John. *Opus Maximum: Or, the Great Essay to Reduce the Moral World from Contingency to System*. London: J. Ginger, 1803.

Thelwall, John. 'A Pedestrian Excursion through Several Parts of England and Wales during the Summer of 1797.' *Monthly Magazine*, VIII (1799) 532–3, 616–19, 783–5, 966–7; IX (1800) 16–18, 228–31.

———. *The Peripatetic; or, Sketches of the Heart, Of Nature and Society; in a series of Politico-Sentimental Journals, in Verse and Prose, of the Eccentric Excursions of Sylvanus Theophrastus*. London: Printed for the Author, 1793.

Thom, W. *Pedestrianism; or, an Account of the Performances of Celebrated Pedestrians during the Last and Present Century*. Aberdeen, 1813.

Thomson, James. *The Seasons*. Ed. James Sambrook. Oxford: Clarendon Press, 1981.

Tillotson, Geoffrey, Paul Fussell and Marshall Waingrow. *Eighteenth-Century English Literature*. New York: Harcourt, Brace & World, 1969.

Walker, Adam. *Remarks Made in a Tour from London to the Lakes of Westmoreland and Cumberland in the Summer of 1791*. London: G. Nicol & C. Dilly, 1792.

Warner, Richard. *A Walk through Wales, in August 1797*. Bath: R. Cruttwell, 1798.

West, Thomas. *A Guide to the Lakes in Cumberland, Westmorland, and Lancashire*. 3rd ed. 1784. Repr. Oxford: Woodstock Books, 1989.
Wilkinson, Joshua Lucock. *The Wanderer; or Anecdotes and Incidents, The Result and Occurrences of a Ramble on Foot, Through France, Germany and Italy, 1791 and 1793*. 2 vols. London: L.J. Higham, 1798.
Wilson, George. *A Sketch of the Life of George Wilson, the Blackheath Pedestrian*. London: the Author, 1815.
Wordsworth, Dorothy. *Journals of Dorothy Wordsworth*. Ed. Mary Moorman. 2nd ed. London: Oxford University Press, 1971.
———. *Journals of Dorothy Wordsworth*. Ed. E. de Selincourt. 2 vols. London: Macmillan, 1952.
Wordsworth, William. *The Borderers*. Ed. Robert Osborn. Ithaca, NY: Cornell University Press, 1982.
———. *Descriptive Sketches*. Ed. Eric Birdsall. Ithaca, NY: Cornell University Press, 1984.
———. *An Evening Walk*. Ed. James Averill. Ithaca, NY: Cornell University Press, 1983.
———. *Letters of William and Dorothy Wordsworth: The Early Years 1787–1805*. Ed. E. de Selincourt. 2nd. ed., revised Chester L. Shaver. Oxford: Clarendon Press, 1967.
———. *Poetical Works*. Ed. E. de Selincourt and Helen Darbishire. 5 vols. Oxford: Clarendon Press, 1940–9.
———. *The Prelude* (1805). Ed. E. de Selincourt, corrected Stephen Gill. London: Oxford University Press, 1970.
———. *'The Ruined Cottage' and 'The Pedlar'*. Ed. James Butler. Ithaca, NY: Cornell University Press, 1979.
———. *The Salisbury Plain Poems*. Ed. Stephen Gill. Ithaca, NY: Cornell University Press, 1975.
———. *William Wordsworth*. Ed. Stephen Gill. Oxford Authors. Oxford: Oxford University Press, 1984.

MAIN SECONDARY SOURCES

Abercrombie, David. *Studies in Phonetics and Linguistics*. London: Oxford University Press, 1965.
Abrams, M.H. *Natural Supernaturalism: Tradition and Revolution in Romantic Literature*. New York: Norton, 1971.
Alexander, Meena. 'Dorothy Wordsworth: The Grounds of Writing.' *Women's Studies*, XIV (1988) 195–210.
———. *Women in Romanticism*. Basingstoke: Macmillan, 1989.
Andrews, Malcolm. *The Search for the Picturesque: Landscape Aesthetics and Tourism in Britain, 1760–1800*. Aldershot: Scolar Press, 1989.
Appleton, Jay. *The Experience of Landscape*. Chichester: John Wiley, 1975.
Appleyard, Donald, Kevin Lynch and John R. Myer. *The View from the Road*. Cambridge, MA: MIT Press, 1964.
Attridge, Derek. *The Rhythms of English Poetry*. London: Longman, 1982.
Aubin, Robert A. *Topographical Poetry in Eighteenth-Century England*. New York: Modern Language Association, 1936.

Bagwell, Philip. *The Transport Revolution from 1770*. London: Batsford, 1974.
Baker, Harold D. 'Landscape as Textual Practice in Coleridge's *Notebooks*.' *ELH*, LIX (1992) 651–70.
Bakhtin, Mikhail. *The Dialogic Imagination: Four Essays*. Trans. Caryl Emerson and Michael Holquist. Austin, TX: University of Texas Press, 1981.
Ball, Patricia. *The Science of Aspects: The Changing Role of Fact in the Work of Coleridge, Ruskin and Hopkins*. London: Athlone Press, 1971.
Barrell, John. *The Idea of Landscape and the Sense of Place, 1730–1840: An Approach to the Poetry of John Clare*. Cambridge: Cambridge University Press, 1972.
Bate, Jonathan. 'The Rights of Nature.' *John Clare Society Journal*, XIV (July 1995) 7–15.
——. *Romantic Ecology: Wordsworth and the Environmental Tradition*. London: Routledge, 1991.
Batten, Charles J. *Pleasurable Instruction: Form and Convention in Eighteenth-Century Travel Literature*. Berkeley and Los Angeles, CA: University of California Press, 1978.
Bayne-Powell, Rosamond. *Travellers in Eighteenth-Century England*. London: John Murray, 1951.
Benjamin, Walter. *Charles Baudelaire: A Lyric Poet in the Era of High Capitalism*. Trans. Harry Zohn. London: New Left Books, 1973.
——. *Illuminations*. Ed. Hannah Arendt. Trans Harry Zohn. London: Jonathan Cape, 1970.
Bennett, Andrew J. '"Devious Feet": Wordsworth and the Scandal of Narrative Form.' *ELH*, LIX (1992) 145–73.
Bermingham, Ann. *Landscape and Ideology: The English Rustic Tradition, 1740–1860*. Berkeley, CA and London: University of California Press, 1986.
Black, Jeremy. *The British Abroad: The Grand Tour in the Eighteenth Century*. Stroud: Alan Sutton, 1992.
Blythe, Ronald. '*Solvitur Ambulando*: John Clare and Footpath Walking.' *John Clare Society Journal*, XIV (July 1995) 17–27.
Bourassa, Steven C. *The Aesthetics of Landscape*. London and New York: Belhaven Press, 1991.
Brownlow, Timothy. *John Clare and Picturesque Landscape*. Oxford: Clarendon Press, 1983.
Buell, Lawrence. *The Environmental Imagination: Thoreau, Nature Writing, and the Formation of American Culture*. Cambridge, MA: Belknap–Harvard University Press, 1995.
Bushaway, Bob. *By Rite: Custom, Ceremony and Community in England 1770–1880*. London: Junction Books, 1982.
Buzard, James T. *The Beaten Track: European Tourism, Literature, and the Ways to 'Culture', 1800–1918*. Oxford: Clarendon Press, 1993.
Chatwin, Bruce. *The Songlines*. London: Jonathan Cape, 1987.
Clifford, James. 'Travelling Cultures.' In *Cultural Studies*. Ed. Lawrence Grossberg, Cary Nelson and Paula A. Treichler. London: Routledge, 1992.
Coe, Charles Norton. *Wordsworth and the Literature of Travel*. New York: Bookman Associates, 1953.

Cohen, Erik. 'A Phenomenology of Tourist Experiences.' *Sociology*, XIII (1979) 179–201.
Cooke, Michael G. 'The Manipulation of Space in Coleridge's Poetry.' In *New Perspectives in Coleridge and Wordsworth*. Ed. Geoffrey Hartman. New York: Columbia University Press, 1972. pp. 165–94.
Copeland, John. *Roads and Their Traffic, 1750–1850*. Newton Abbot: David & Charles, 1968.
Cox, Jeffrey N. 'Keats in the Cockney School.' *Romanticism*, II (1996) 27–39.
Daniels, Stephen, and Charles Watkins. *The Picturesque Landscape*. Nottingham: Department of Geography, University of Nottingham, 1994.
Dyos, H.J., and D.H. Aldcroft. *British Transport: An Economic Survey from the Seventeenth Century to the Twentieth*. Harmondsworth: Penguin, 1974.
Elder, John. *Imagining the Earth: Poetry and the Vision of Nature*. Urbana and Chicago, IL: University of Illinois Press, 1985.
Everest, Kelvin. *Coleridge's Secret Ministry: The Context of the Conversation Poems 1795–1798*. Hassocks, Sussex: Harvester Press, 1979.
Fish, Stanley. *Doing What Comes Naturally: Change, Rhetoric, and the Practice of Theory in Literary and Legal Studies*. Oxford: Clarendon Press, 1989.
Foster, Shirley. *Across New Worlds: Nineteenth-Century Women Travellers and their Writings*. Hemel Hempstead: Harvester Wheatsheaf, 1990.
Gilbert, Roger. *Walks in the World: Representation and Experience in Modern American Poetry*. Princeton, NJ: Princeton University Press, 1991.
Hankinson, Alan. *Coleridge Walks the Fells: A Lakeland Journey Retraced*. Maryport, Cumbria: Ellenbank Press, 1991.
Harding, D.W. *Words into Rhythm: English Speech Rhythm in Verse and Prose*. Cambridge: Cambridge University Press, 1976.
Hartman, Geoffrey H. *Wordsworth's Poetry 1787–1814*. New Haven, CT and London: Yale University Press, 1964.
Haughton, Hugh, Adam Phillips and Geoffrey Summerfield (eds). *John Clare in Context*. Cambridge: Cambridge University Press, 1994.
Hayden, Donald E. *Wordsworth's Walking Tour of 1790*. Tulsa, OK: University of Tulsa Press, 1983.
Hayden, J.O. *Romantic Bards and British Reviewers*. London: Routledge & Kegan Paul, 1971.
Heaney, Seamus. *Preoccupations: Selected Prose, 1968–1978*. London: Faber, 1980.
Hibbert, Christopher. *The Grand Tour*. London: Weidenfeld, 1969.
Holmes, Richard. *Coleridge: Early Visions*. London: Hodder & Stoughton, 1989.
Hussey, Christopher. *The Picturesque: Studies in a Point of View*. 1927. Repr. London: Frank Cass, 1967.
Izenberg, Gerald N. *Impossible Individuality: Romanticism, Revolution, and the Origins of Modern Selfhood, 1787– 1802*. Princeton, NJ: Princeton University Press, 1992.
Johnston, Kenneth R. *Wordsworth and 'The Recluse'*. New Haven, CT and London: Yale University Press, 1984.
Keach, William. 'Cockney Couplets: Keats and the Politics of Style.' *Studies in Romanticism*, XXV (1986) 182–96.

Kelley, Theresa M. *Wordsworth's Revisionary Aesthetics*. Cambridge: Cambridge University Press, 1988.
Knight, Frida. *University Rebel: The Life of William Frend (1757–1841)*. London: Victor Gollancz, 1971.
Lambert, R. S. *Grand Tour: A Journey in the Tracks of the Age of Aristocracy*. London: Faber, 1935.
Leed, Eric J. *The Mind of the Traveller: From Gilgamesh to Global Tourism*. New York: Basic Books, 1991.
Levin, Susan M. *Dorothy Wordsworth and Romanticism*. New Brunswick, NJ and London: Rutgers University Press, 1987.
Liu, Alan. *Wordsworth: The Sense of History*. Stanford, CA: Stanford University Press, 1989.
Long, Richard. *Richard Long: Walking in Circles*. London: Thames & Hudson, 1991.
Marples, Morris. *Shank's Pony: A Study of Walking*. London: Dent, 1959.
McCormick, Anita Hemphill. '"I shall be beloved – I want no more": Dorothy Wordsworth's Rhetoric and the Appeal to Feeling in *The Grasmere Journals*.' *Philological Quarterly*, LXIX (1990) 471–93.
McGann, Jerome. *The Textual Condition*. Princeton, NJ: Princeton University Press, 1991.
McKusick, James. 'From Coleridge to John Muir: The Romantic Origins of Environmentalism.' *The Wordsworth Circle*, XXVI (1995) 36–40.
Mellor, Anne K. 'Coleridge's "This Lime-Tree Bower My Prison" and the Categories of English Landscape.' *Studies in Romanticism*, XVIII (1979) 253–70.
——. *Romanticism and Gender*. New York and London: Routledge, 1993.
Michasiw, Kim Ian. 'Nine Revisionist Theses on the Picturesque.' *Representations* 38 (Spring 1992) 76–100.
Modiano, Raimonda. *Coleridge and the Concept of Nature*. London and Basingstoke: Macmillan, 1985.
Moir, Esther. *The Discovery of Britain: The English Tourists 1540–1840*. London: Routledge, 1964.
Murphy, Dervla. *Where the Indus is Young: Walking in Baltistan*. 1977. Repr. London: Arrow, 1991.
Nabholtz, John R. 'Dorothy Wordsworth and the Picturesque.' *Studies in Romanticism*, III (1964) 118–28.
Ousby, Ian. *The Englishman's England: Taste, Travel and the Rise of Tourism*. Cambridge: Cambridge University Press, 1990.
Price, Martin. 'The Picturesque Moment.' In *From Sensibility to Romanticism: Essays Presented to Frederick A. Pottle*. Ed. F.W. Hilles and Harold Bloom. New York: Oxford University Press, 1965. pp. 259–92.
Robinson, Jane. *Wayward Women: A Guide to Women Travellers*. Oxford: Oxford University Press, 1990.
Robinson, Jeffrey. *The Walk: Notes on a Romantic Image*. Norman, OK and London: University of Oklahoma Press, 1989.
Roe, Nicholas. *The Politics of Nature: Wordsworth and Some Contemporaries*. Basingstoke: Macmillan, 1992.
Rookmaaker, H.R. *Towards a Romantic Conception of Nature: Coleridge's Poetry up to 1803*. Amsterdam: John Benjamins, 1984.

Rossi, William. '"The Limits of an Afternoon Walk": Coleridgean Polarity in Thoreau's "Walking".' *ESQ*, XXXIII, ii (1987) 94–109.
Ruddick, William. '"As much diversity as the heart that trembles": Coleridge's Notes on the Lakeland Fells.' In *Coleridge's Imagination: Essays in Memory of Pete Laver*. Ed. Richard Gravil, Lucy Newlyn and Nicholas Roe. Cambridge: Cambridge University Press, 1985. pp. 88–101.
Schor, Naomi. *Reading in Detail: Aesthetics and the Feminine*. Methuen: New York and London, 1987.
Simpson, David. *Wordsworth's Historical Imagination: The Poetry of Displacement*. London: Methuen, 1987.
Sperry, Stuart. *Keats the Poet*. Princeton, NJ: Princeton University Press, 1973.
Taplin, Kim. *The English Path*. Woodbridge, Suffolk: Boydell Press, 1979.
Towner, John. 'The Grand Tour: A Key Phase in the History of Tourism.' *Annals of Tourism Research*, XII (1985) 297– 333.
Turner, Victor. *The Ritual Process: Structure and Anti-Structure*. London: Routledge, 1969.
Urry, John. *The Tourist Gaze: Leisure and Travel in Contemporary Societies*. London: Sage, 1990.
Walker, Carol Kyros. *Walking North with Keats*. New Haven, CT and London: Yale University Press, 1992.
Wallace, Anne D. *Walking, Literature, and English Culture: The Origins and Uses of Peripatetic in the Nineteenth Century*. Oxford: Clarendon Press, 1993.
Watson, J.R. *Picturesque Landscape and English Romantic Poetry*. London: Hutchinson, 1970.
Williams, Raymond. *The Country and the City*. London: Chatto & Windus, 1973.
Woolsey, Linda Mills. 'Houseless Woman and Travelling Lass: Mobility in Dorothy Wordsworth's *Grasmere Journals*.' *Tennessee Philological Bulletin*, XXVII (1990) 31–8.

Index

Abercrombie, David, 85
Abrams, M. H., 116
Aikin, Arthur, 11
Alexander, Meena, 162, 163, 174–5
Appleton, Jay, 143–5
Appleyard, Donald, 136–7
Attridge, Derek, 86–7
Aubin, R. A., 92

Bagwell, Philip, 20–1
Baker, Harold, 130, 135
Bakhtin, Mikhail, 202–3
Ball, Patricia, 130, 133
Ballantyne, Andrew, 222 n
Barclay, Captain, 3, 4, 22
Barrell, John, 73, 161, 179, 183, 184–5, 186–7, 188
Bate, Jonathan, 187–8
Batten, Charles, 46–7
Bayne-Powell, Rosamond, 218 n
Benjamin, Walter, 207, 210–11
Bennett, Andrew, 224 n, 225 n
Bewell, Alan, 176
Bingley, Rev. William, 11–12, 29, 43–4
Black, Jeremy, 15, 16
blank verse, 83, 84–8, 109, 139
Blythe, Ronald, 160, 177–8
Bourassa, Steven, 145, 228 n
Bowen, Miss M., 83–4
Bowles, William Lisle, 7, 9, 78–9
Bristed, John, 12, 26
Brown, Charles Armitage, 196, 197–8, 206
Brownlow, Timothy, 185
Budworth, Joseph, 9
Buell, Lawrence, 189
Burke, Edmund, 57–9, 129
Bushaway, Bob, 178–9, 188
Butler, Marilyn, 201
Buzard, James, 51–2

Chatwin, Bruce, 69–70
Chisholm, A. R., 85

Clare, John, 159–62, 176–91
  WORKS: 'Autobiography', 177, 178, 190; 'Emmonsales Heath', 190–1; 'Emmonsails Heath in Winter', 186; 'Evening', 177; 'The Fens', 182; 'The Flitting', 112, 189–90; 'I Am', 177; 'Journey out of Essex', 159–61; 'The Lamentations of Round-Oak Waters', 181–2; 'Letter to The Every-day Book', 179; 'The Mores', 182–4; 'Out of Door Pleasures', 176, 186; 'Reccolections after a Ramble', 185–6; 'The Robin's Nest', 187; 'The Squirrel's Nest', 187; 'Sunday Walks', 179–81
Clifford, James, 158, 161–2
Cockney School, 199–201, 205–6, 210
Coe, Charles Norton, 220–1 n
Cohen, Erik, 41–2
Coleridge, Samuel Taylor, 10–11, 34–5, 90, 119, 126–54, 198
  WORKS: 'The Ancient Mariner', 127; 'The Aolian Harp', 143; 'Fears in Solitude', 143, 146–8; 'Inscription for a Seat by the Road Side', 139–40; 'Kubla Khan', 127; 'This Lime-Tree Bower My Prison', 143, 148–50; 'Lines Composed While Climbing the Left Ascent of Brockley Coomb', 143; 'Lines Written in the Album at Elbingerode', 150–1; *Notebooks*, 130–7; 'Perspiration', 34–5, 129–30, 139; 'The Picture', 151–3; 'Reflections on Having Left a Place of Retirement', 143,

Coleridge (*contd.*)
  145–6; 'A Stranger Minstrel', 140–1; 'The Wanderings of Cain', 144–5; 'To a Young Friend', 141–3
Combe, William, 55
Cooke, Michael, 134, 144, 146
Cottle, Joseph, 127
Cowper, William, 79–83
Cox, Jeffrey N., 199–200, 201, 206
Coxe, William, 6–7, 9
Crowe, William, 34, 39
  Lewesdon Hill, 71, 75–7

Denham, Sir John, 71–2
De Quincey, Thomas, 8, 25, 89
  Confessions of an English Opium-Eater, 209–10
Dyer, George, 11
Dyer, John, 71, 72–5
Dyos and Aldcroft, 20, 21

Elder, John, 114, 115–16, 122
enclosure, 181–4
Everest, Kelvin, 146, 148

Fillippi, Fillippo de, 69
Fish, Stanley, 52
flâneur, 209, 210
Foster, Shirley, 158, 229 n
Frend, William, 7–8, 9, 34, 39
Freud, Sigmund, 133

Gay, John, 207–8
Gilbert, Roger, 184
Gilpin, William, 9, 50, 53, 54–5, 56, 57, 58–9, 61, 62–7, 110
Gittings, Robert, 197
Goldsmith, Oliver, 77–8
Goodridge, John, 181
Gray, Thomas, 50
Greever, Garland, 217 n

Hankinson, Alan, 226 n
Harding, D. W., 85–6
Hartman, Geoffrey, 122, 123, 124
Hayden, Donald E., 224 n
Hazlitt, Sarah, 155–8

Hazlitt, William, 14, 90, 126, 134, 155–6, 196, 197
  Liber Amoris, 208–9; 'On Going a Journey', 192–5, 211
Heaney, Seamus, 90–1, 124, 184
Hibbert, Christopher, 218 n
Holmes, Richard, 226 n, 227 n
Hucks, Joseph, 10–11, 25, 29, 34–5, 39, 42, 44, 128, 130
Hunt, Leigh, 22, 195, 199, 210–15
  'A Rustic Walk and Dinner', 213–15; 'Walks Home by Night', 212–13
Hussey, Christopher, 56, 62–3
Hutton, William, 12, 20, 49

Izenberg, Gerald, 98, 100, 225 n

Jakobson, Roman, 46
Jeffrey, Francis, 206
Johnston, Kenneth, 115, 117, 121, 122

Keach, William, 200, 206
Keats, John, 195–206
Kelley, Theresa, 105, 106
Knight, Richard Payne, 53, 54–5, 56–7

Lambert, R. S., 218 n
Leed, Eric, 23, 27–8, 32–3, 39, 40, 45–6, 68, 120–1, 159
Levin, Susan, 163, 165, 170, 174
Lindenberger, Herbert, 226 n
Liu, Alan, 54, 92
Long, Richard, 68–9

Marples, Morris, 5, 6, 12
McCormick, Anita, 172, 230 n
McGann, Jerome, 198
McKusick, James, 178
Mellor, Anne, 148–9, 166, 228 n
Michasiw, Kim Ian, 54, 55, 63–4
Modern Universal British Traveller, 46
Modiano, Raimonda, 130–1, 133, 138, 145, 148, 151, 153
Moritz, Carl, 9, 24–5

Nabholtz, John, 173
Newell, Robert, 13
Nugent, Thomas, 15

Observant Pedestrian, The, 1, 11
Ousby, Ian, 221 n

pedestrianism
  aesthetic, 49–53; athletic, 1, 2–4, 11; and class, 22–7, 33–9, 155, 159–62, 167, 168–9, 176–8, 179–80; Continental, 6–9, 99; and ecology, 69, 143, 148, 151, 187–91; etymology of, 1–2; and gender, 155–9, 164, 165–6, 172–6; ideology of, 27–8, 30, 211; in Lake District, 9–10; and literary form, 32–3, 62–88, 90–1, 115–16, 118–20, 124–5, 138, 149–50, 166, 184–6; mental characteristics of, 67–70, 120–1, 132–7, 148, 193, 208–9; philosophical, 42–8; and pilgrimage, 39–42; radical, 33–9, 97, 128–30, 162, 169; as rite of passage, 37–9; in Scotland, 26, 156–8, 172–4, 175–6, 195–206; textual, 91, 93, 96, 101, 104, 125; urban, 195, 206–13; in Wales, 10–12, 25, 26, 128–30
Penhouet, Mandet de, 11, 29
perambulations, parish, 178–9, 188
Peripatetic The, 30–2, 33, 44–5
picturesque, 53–61, 62–7, 92–6, 97–8, 109–10, 131–2; and mental play, 62–7; politics of, 54–5, 63–4; rhetoric of, 55–61; travel, 49, 53, 55, 60–1, 66–7
Plumptre, James, 12
Poe, Edgar Allan, 209
Porter, Roy, 159–60
Powell, Foster, 3, 7, 9, 10–11, 22
Price, Uvedale, 53, 54–5, 57, 59–60, 114

Radcliffe, Ann, 158
Recollections of a Pedestrian, 30
Reed, Mark, 122
Robinson, Jane, 158
Robinson, Jeffrey, 48, 212

Roe, Nicholas, 226n
Rookmaaker, H. R., 153
Rossi, William, 228n
Rousseau, Jean-Jacques, 31, 32
Ruddick, Bill, 130, 131

Santayana, George, 70–1
Shelley, Mary, 158
Simpson, David, 108, 115
Skrine, Henry, 9–10
Southey, Robert, 10, 34, 39–40, 50, 127, 129, 130
Sperry, Stuart, 196, 203
Stansbury, P., 26–7
Stewart, John 'Walking', 8, 9
Swift, Jonathan, 207

Taplin, Kim, 93
Taylor, John, 184
Thelwall, John, 12, 21, 22, 25–6, 35–7, 39
Thom, Walter, 2–4
Thompson, E. P., 220 n
Thomson, James, 74
Thornton, Kelsey, 181
topographical poetry, 71–7, 92
Towner, John, 16
Traherne, Thomas, 125
travel
  conditions of, 19–22; cost of, 21; Grand Tour, 14–17, 42, 77, 159; mental effects of, 45–6, 66–7; philosophical, 45–8, 77–8; relation to tourism, 40–2, 168; speed, 21–2, 136–7 *see also* pedestrianism
Turner, Victor, 37–8

Urry, John, 40–1

vagrancy, 81, 107–8, 109, 111, 178
Valéry, Paul, 85

Walker, Adam, 9, 29, 44, 49–51, 52
Walker, Carol Kyros, 198, 203, 231–2 n
walking, *see* pedestrianism
Wallace, Anne, 5–6, 9, 18–19, 21, 22, 23, 32, 66, 79, 80, 89–90, 113, 114, 115, 117, 176

Warner, Richard, 11, 12, 25
Watson, J. R., 92
West, Thomas, 9, 20, 50, 56
Wilkinson, Joshua, 8
Williams, Helen Maria, 158
Williams, Raymond, 206–7, 208, 209
Wilson, George, 3, 12
Wollstonecraft, Mary, 158
Woolsey, Linda Mills, 172
Wordsworth, Dorothy, 97, 101, 127, 133, 158, 162–76, 195
  WORKS: *Alfoxden Journal*, 163; 'Grasmere – A Fragment', 164–5; *Grasmere Journal*, 163–4, 165–7, 172; *Journal of a Tour on the Continent*, 167–72, 174–5; *Recollections of a Tour Made in Scotland*, 172–4, 175–6

Wordsworth, William, 1, 10, 22, 32, 35, 40, 42, 60–1, 66, 87, 89–125, 126, 127, 144, 193, 208, 209
  WORKS: *Adventures on Salisbury Plain*, 106–8; *The Borderers*, 108–9; 'Composed upon Westminster Bridge', 212; *Descriptive Sketches*, 97–101, 102, 116, 124; *An Evening Walk*, 91–6, 111; *Home at Grasmere*, 112–14, 124, 165; 'Michael', 90; *The Prelude*, 101, 102–3, 104–6, 114–25, 195; *The Ruined Cottage*, 109–12, 124; *Salisbury Plain*, 102–4; 'A Slumber did my Spirit Seal', 114; 'Tintern Abbey', 193; 'When first I journeyed hither', 90
Wright, A. R., 231 n